THE SKY IS FALLING!

THE SKY IS FALLING!

Why Buildings Fail

Marvin Hornstein

South Bend, Indiana

THE SKY IF FALLING
Why Buildings Fail

and books
702 South Michigan, South Bend, Indiana 46618

Library of Congress Catalog Number: 82-072602

International Standard Book Number: 0-89708-106-4

The publisher is grateful for assistance
provided by Thomas Lyon, architect.

Printed in the United States of America

Additional copies available:
the distributors
702 South Michigan
South Bend, IN 46618

This book is respectfully dedicated
to the construction industry,
for better or worse.
Without it, we might all be watching
our television sets
while sitting in our tents.

Table of Contents

INTRODUCTION

Have you stood in awe of the megaliths that surround and comprise our modern cities; that spread over suburban shopping centers; are the pride of our college campuses; and offer shelters as civic sports arenas and even churches? These manmade mountains of concrete, steel and glass often appear to mute ancient wonders as they massively rest, and seemingly broadcast a sense of structural eternity. The pride of builders, these monuments to our architectual ingenuity offer great spaces in which to carry out our daily work and play. Along with their magnificence these buildings also communicate a sense of stability; a reassurance that future generations will surely admire and appreciate our industry. Unfortunately this reassurance may all be an illusion, a false security, perhaps as other past dreams of structural immortality have been, from the Gardens of Babylon to the recent catastrophes of massive building failures in our own times.

This study should be of concern to anyone who enters a modern building. Included are homes, apartments, schools, and office buildings, as well as shopping centers, supermarkets, bowling alleys, and restaurants. The list continues with auditoriums, theaters, hotels, gymnasiums, churches, synagogues, parking structures... and any other type of building that we may think of.

The key words are: public safety. There is today a commonly shared concern by builders, architects and engineers that most people know very little of the risks they take daily in entering or occupying many buildings. The situation is worsened by a broad-based unawareness, and lack of interest especially in buildings with a long extended roof span. I am specifically concerned because we all have the right to expect personal safety when we enter any building.

In recent years much attention has been given to many forms of safety; construction, housing, vehicle, consumer product, and protection of workers in regard to pollution and industrial accidents — the lists of safety rules and recommendations seem endless. And yet, when we hear of a hotel walkway falling, or an arena roof collapsing from the weight of snow or water, we all feel helpless. We are beyond our day-to-day expectations and chalk it up to "acts of fate." Worse yet, we distribute the blame to everyone that may have been even marginally involved with the disaster. In many cases we incorrectly conclude that an appeal for everyone to "do a better job" will somehow transform the building industry's practices and thereby make all future buildings suddenly safe. This approach may temporarily silence the problem, and may even give psychological comfort, but it is again shown inadequate when another building collapses, and our attention is abruptly re-focused by the tragic consequences. Breaking this cycle of wishful-building is no small matter. It may already be so fixed in the workings of our economy that we shall be stuck with it for a long time to come.

The situation is far worse than we might realize. Few people are even aware of the vast numbers of actual structural building failures that have occurred. In recent years alone the financial losses have run into the billions of dollars. The number of deaths and injuries is staggering and many of these building failures occur before the buildings are even completed. Generally, unless it is a spectacular collapse causing many fatalities, the losses rarely receive widespread publicity.

The main purpose for this book is to enlighten the general public regarding safety issues in building construction. To do so I will take you behind the scenes of building design and its execution. I have purposely avoided the overly technical aspects involved, but I reveal how economic and political powers combine from the inception of a building project resulting in unsafe designs and hazardous structures.

The following chapters explore how the decision makers, those who hold power in the construction industry, often fail to comprehend the great degree of public danger and the risk they cause. Rising interest rates have also added to this safety problem, along with meager and unfair salaries paid to members of the architectural and engineering professions. These elements, combined with general apathy and the slick public relations of building owners, have in many cases led the public to become the unsuspecting "guinea pigs," and ultimate victims, of the deficiencies within today's building trade. This study also offers some positive steps that may be taken to increase the safety of all structures, since there is absolutely no reason why the present condition cannot be dramatically improved with the reasonable efforts of those engaged in the building trade.

Although this subject is a widespread concern, it should especially be of interest to the members of the building industry, from draftsmen to architects, engineers to contractors, and building inspectors to developers. Those who work on the inside should perhaps best recognize and understand many of the problems that are here discussed. In particular, I am anxious that the issues of faulty building design and construction find their way into our educational system. Although the universities do an excellent job in the technical and social training of future building professionals, they often fail to highlight the "real" world in which all students will eventually work. Often, the shock of a low salary and very little hope of financial advancement, combined with the awesome responsibility converge upon the neophyte. Newcomers to the profession often begin to look for moonlighting work very early in their careers. Others, likewise competently trained, quickly realize the building industry's unrealistic expectations and change their profession. Unfortunately, the truth is often painful. But the concealment of the truth may eventually lead to a greater harm. It is evident that our educational system has failed to raise its student consciousness to the many shortcomings; not only within the design profession, but of the entire construction industry. These issues must now be faced, otherwise the final consequence of this continued academic blindness to the practical problems faced by graduates will the eventual loss of many qualified and creative practitioners.

Again, this matter concerns us all — our loved ones, relatives, and friends. It also concerns the largest single industry in our country: building construction. If this study can call attention to the need for more public information and lead to accountability in this immense industry, if it can perhaps help to save one life by preventing one building collapse, then the effort is well worth the attention.

Photograph on preceeding page:

July, 1979. Emergency crews gather around the Horizon sports stadium in Rosemont, Illinois after the roof collapsed during construction. Damage estimates totaled 3 million dollars.

—*Chicago Tribune photo. Used with permission.*

Chapter One

WHY SO MANY BUILDING FAILURES?

When a medical doctor makes a serious mistake, the patient, often one person, may die. Yet when a structural engineer makes a crucial error, many, perhaps hundreds of people can be injured or killed. The doctor is generally backed by an elaborate support system which often allows time for correction of errors in judgement through either a change in the method of treatment or by the use of different drugs. But if a single important beam, bolt, or washer fails in a building structure, there is rarely time for remedial review. Most often it is too late.

The medical practioner is usually afforded the resources to act in the best interests of the patient, and normally, a doctor will personally handle the care of that patient. The art of healing is the result of many years of medical school and hospital training through internship and residency programs, which all work to ensure that medical care is properly performed. The professional, consciencious doctor is rarely compromised or intimidated into accepting either a low fee for medical services or an unreasonable time schedule. Nearly everyone recognizes the importance and the nature of medical work.

In contrast, although the building industry shares an internal level of professional respect for its responsibility to the public welfare; members of this industry rarely enjoy the same recognition as those in the medical field. Somehow, the immediacy, or visible contact of architects and engineers with the erected structure is distant and often assumed to be wasteful. Thus, architects and structural engineers are frequently compelled to hire less qualified individuals than themselves to attend to the actual work of design, such as the tedious engineering calculations and critical drawings of details and plans. Each brokered or sub-contracted responsibility holds the potential for causing damage, injury, or even death when and if inaccuracies occur in judgment are made at these planning stages of building design. Yet, the architect and engineer are frequently rushed into a schedule that allows too little time for a flawless job, and even less time for checking or reviewing the work at its most primary planning stage. In the smallest concrete building project, with tens of thousands of structural members and connections, the need for close scrutiny can never be overemphasized. One small error could mean disaster.

The pressures associated with scheduling and coordinating building construction are immense. Owners, builders, architects, engineers and developers are each expected to prepare their plans rapidly and often as cheaply as possible. The tension is further intensified by low engineering and architectural fees and the overall budgets which ultimately control the expense of final construction. As a result of rising labor and material costs during recent years, the trend in the building industry has been to direct expenses to external appearance, such as elaborate finishes and public decoration. Of course the imbalance of moneys spent has lead many builders to skimp on unseen items such as covered-over structural members.

The phenomenal advances in recent technology have permitted the use of smaller structural members in many buildings, including shallower beams with the use of higher strength steels, and smaller concrete sections due to the development and use of prestressed concrete and sophisticated computerized design methods. Yet, it remains a fact that while the strengths in materials have dramatically increased, there has been no comparable increase in the overall stiffness of these materials. For example, today a 12-inch deep steel beam may be specified for use in a floor structure system. Had the same building been designed in the 1950s, chances are that a 14 or 16-inch deep beam may have been called for, thereby offering a much stiffer floor system. Without giving sufficient thought to potential problems, some designers today are responsible for "bouncy" floors and sagging roofs.

It is also well recognized throughout the American building industry that whenever expense budgets are planned for a new building, the raw material expenses for construction fall secondary to the human values of the labor required to assemble these materials. This view is most likely a result of our economic system and forms the foundation of a commonly held belief about the overall values for various types of labor performed by the construction industry.

The low fees paid to architects and structural engineers are probably the most direct cause for many of the problems we face today. We have all heard the expression: "you get what you pay for!" It is obvious that if you hire a $35 per hour attorney to handle a lawsuit, you should feel no right to expect the same results as you might from an attorney who charges $125 per hour. Likewise, this is the bottom line in the building industry, and often at the root of most construction problems.

Assume that a finished plan is accurately prepared by an architect and engineer, and then passed on to a contractor for execution. What assurance is there that the building will be constructed in keeping with the original plans? Surprising to many, there is effectively no mechanism for checking the plans against the actual work done! In many cases there is little, if any, field supervision by an architect or structural engineer. Buildings, like all property, are the ultimate responsibility of the owner who meets the particular standards of a State and from there on is free to call all the shots. The owner defines what emphasis is worth the cost and has the final word with all who work on the building project. The architect and engineer are rarely extended the responsibility, or privelege, to oversee the actual work on a project beyond the clarification of their plans. Once completed, building plans become the full property of the owner who assigns the responsibility of their execution to a construction firm. If an owner, whimsically decides to spend most of the construction budget on finishing materials, or colorful facades and leaves much of the structural work inspection to a third-party testing laboratory, or the state building inspector; the original designer or architect can do little more than join the thousands of pedestrians who squint through panelled fences as the building is erected. Many past structural failures have been shown to be the result of inadequate supervision and inspection, even though the original plans were found to be flawless.

An example of this occurred several years ago when my office executed the structural design for a restaurant in Omaha, Nebraska. We used prefabricated wood trusses for its roof structure. Shortly after the roof was completed I visited the site and was shocked to find that most of the trusses were far too short to have full bearing on top of the supporting walls. The contractor had simply nailed a small wooden ledger strip to the sides of the walls in order to support the shorter trusses. On checking the capacity of the nails and ledger, I found that they would most certainly have failed during the first heavy snowstorm and may have brought the entire roof structure down. Fortunately, I was able to correct the error in time. But what would have happened had I not visited the project at that time? It is doubtful that the true reason for that roof failure would have ever been found beneath the heap of snow-covered rubble.

Occasionally, the architect and engineer may decide to use a relatively new and perhaps unproven engineering technique which seems acceptable in theory. In these cases, the general public often becomes the "test group" of such experimental designs. Although a costly procedure, these types of buildings should be load-tested during their construction to assure safety *prior* to public use. One example of this requisite occurred recently when after a series of large auditorium roof collapses, President Gerald Ford insisted that load-testing procedures be used during the building of the Museum bearing his name at Grand Rapids, Michigan. He thereby assured the adequacy of its space-frame roof. I feel that this kind of additional expense is money well spent.

The fierce competition from many giant aerospace and defense-oriented companies, who often can afford to pay higher salaries for qualified engineers, has made it increasingly difficult for many architectural and consulting engineering firms to hire top-notch emloyees for work in building design. Since low fees have become the standard for the industry, these firms simply cannot afford to hire qualified workers. The reality is that if the smaller independent firm is to stay in business, its working staff must operate within a fixed salary ceiling.

In summary, a comprehensive list of causes for building failures would be far too detailed to exactly account in any single book of this length. Yet at the risk of over simplifying the problem, it is quite clear that the greatest contributing factors are the lack of sufficient *time* and *money*. Hurried and unrealistic building schedules, combined with a disproportionate emphasis on appearance are responsible for most actual or near errors on the plans, and low fees to designers and engineers only assure that these errors will continue to go unnoticed and uncorrected.

The following list of structural failures was extracted from reports in national engineering publications. It covers the most recent six year period, and is intended here only as a representative sample of the types of failures that occur with increasing regularity.

ACTUAL STRUCTURAL FAILURES IN RECENT YEARS

This is only a partial list representing typical examples of actual failures during this period.

March, 1976/ SUSPENDED CEILING COLLAPSES

A suspended plaster ceiling over an indoor swimming pool in Long Beach, California, collapsed and killed one person and injured four others. A 20 x 40 foot section fell into the pool, shoving the water in a wave which shattered the glass door entrance to the hall. The person killed was trapped under the weight of the material and drowned.

March, 1976/ OLYMPIC STADIUM PIECES FALL, KILL FOUR

In the second fatal accident at the Montreal Olympics site, four workers died when two, twenty ton precast sections fell 135 feet from the main stadium.

April 1976/ HANGAR ROOF COLLAPSES

A cantilevered hangar roof under construction at North Island Naval Air Station in San Diego, California, collapsed injuring five workers. The roof, 10 x 576 feet, 35 feet high, gave way while workers were leveling it before making connections.

January, 1977/ CALIFORNIA COURTS BUILDING SETTLES

The 8-year old Orange County Courts Building, which had settled up to 5 inches, required an $800,000 repair job.

February, 1977/ MARBLE PANELS FALL FROM NEW YORK SKYSCRAPER

Stresses created by wind caused several 100 pound marble panels to fall from an 11-year old New York City skyscraper.

March, 1977/ JFK CENTER DAMAGE IS $4.5 MILLION

Water leaking into the John F. Kennedy Center for the Performing Arts in Washington, D.C., caused 4.5 million dollars in damage to the $73 million building.

May, 1977/ WINDOW WALL BLOWS OUT

A 60 foot high window wall in a state office building in Hauppauge, New York, blew out during a storm. An investigation was begun.

September, 1977/ U.S. SUES OVER SINKING BUILDING

The U.S. Justice Department filed a 6 million dollar suit against four contractors of the 85 million dollar Federal Courthouse and Office Building in downtown Philadelphia, Pennsylvania. The charges were for negligence in the construction of the building's foundations, causing the building to sink.

May, 1977/ HANCOCK TOWER LEGAL DEBATE

All 10,344 panes of exterior reflective glass of the 60-story John Hancock Tower in Boston, Massachusetts, had to be replaced after breakage that began in November of 1972. The building's cost increased from $112 million to $144 million. Also some 1,500 tons of structural steel had to be added to brace the building. Occupancy was delayed by more than 3 years. The cause of the failure and who was to blame are still the subject of court papers and an extensive legal debate which is expected to last several years.

January, 1978/ SPACE-FRAME ROOFS COLLAPSE

A 2.5 acre space-frame roof of the Civic Center Coliseum in Hartford, Connecticut, collapsed just before dawn. Heavy equipment sifted through the wreckage as experts searched for clues of what caused the failure of the five year old structure. The failure began when the main center portion of the roof fell 83 feet to the 12,000 seat arena below. No one was injured, since the arena was not in use at the time. Design flaws were suspected. Other roofs collapsed in two other Connecticut towns, and one in West Virginia.

Another space-frame roof, a 171 foot diameter dome over a 3,500 seat auditorium, collapsed at a college in Brookville, New York. No one was injured.

February, 1978/ POST OFFICE AND TV TOWER COLLAPSE

A portion of a post office in Massachusetts and a 790 foot television tower in Kansas were two collapses from the winter's storms. Damage was estimated at over one million dollars.

March, 1978/ CONCRETE ROOF COLLAPSES, KILLS TWO

An investigation was begun into the collapse of seven 23 x 50 foot sections of a vaulted prestressed concrete roof under construction for a subway maintenance job in Rio de Janeiro. Two workers were killed, and 19 injured.

May, 1978/ FORM COLLAPSE KILLS 51 ON COOLING TOWER

A formwork system atop a power plant cooling tower in West Virginia collapsed, sending 51 workers plummetting 168 feet to their death in one of the world's worst construction tragedies.

May, 1978/ DESIGNER BLAMED IN CRACKED GLASS CASE

In what was to be a precedent for material designers and manufacturers, a jury in Indianapolis, Indiana, awarded several million dollars in damages to the owner of three 11-story office buildings plagued with problems of cracking and leaking windows.

June, 1978/ ENGINEERS STUDY FLAT-ROOF COLLAPSE

Engineers in Garland, Texas, investigated the collapse of a church roof that killed a 9 year old girl and injured 57 persons. City officials claim the structure complied with building codes and state law. The failure occurred during morning services as tons of water from an overnight rainstorm fell on worshippers below. The building was 2.5 years old.

June, 1978/ TOWER COLLAPSE KILLS TWO

The partial collapse of a cooling tower under construction for a Commonwealth Edison Company Nuclear Power Plant in Illinois may have been caused by temporary steel supports. Two workers died, and two were injured.

August, 1978/ 25 MILLION DOLLAR TENNESSEE ROOF SAGS

Investigators reviewed the possible causes of a sagging roof at a City-County building in Tennessee.

September, 1978/ WASHINGTON TRANSIT LINE ROOF SAGS

A design flaw may have caused a roof deflection problem in a Washington, D.C., Transit Line building, according to engineers. The station was not yet in use.

December, 1978/ CODES CALLED INADEQUATE

Concern grew in the U.S. over the lack of adequate criteria for design of large panel precast structures to protect against progressive collapse. Local failures appear to spread gradually and result in a major or total collapse.

December, 1978/ MARBLE PANELS FALL FROM HIGH RISE

Two marble panels, weighing 600 pounds each, fell from the face of a New York City department store. No one was injured. The street was temporarily closed for an investigation into the failure.

December, 1978/ 2,000 TON CONCRETE SLAB COLLAPSES

A prestressed concrete platform collapsed in Rio de Janeiro. Six visitors were killed at the Exposition and Convention Center.

January, 1979/ CITY-COUNTY BUILDING NEEDS $1,000,000 IN STRUCTURAL REPAIRS

Faulty beam and column joints were discovered after a weld failed and a portion of the structure sank two inches.

January, 1979/ MORE THAN 100 ROOFS FALL IN CHICAGO

When snow depths reached 30 inches, scores of roofs collapsed throughout the Chicago area.

March, 1979/ 4.5 TON SLAB FALLS FROM GYM

A large precast concrete panel fell two stories onto a walkway last week. No one was injured.

March, 1979/ DESIGN DISPUTE HALTS SPACE FRAME JOB

The City of Detroit halted a major project at a downtown mall due to the discovery of inadequate foundations and columns already in place.

April, 1979/ MARBLE PANELS FALL

Early this month, six marble panels fell off the Chicago Water Tower Place. In Albany, New York, similar failures occurred.

May, 1979/ HARTFORD COLISEUM REBUILT

Sixteen months after the 360 x 300 foot space truss roof gave way and collapsed, steel trusses were taking shape to hold a larger roof area for 14,500 spectators.

June, 1979/ ARENA ROOF FALLS IN KANSAS CITY

The 17,000 seat Kemper Arena in Kansas City, Missouri, was the site of a major collapse in which the roof structure crashed 95 feet to the floor below during a rainstorm. Two nights earlier, 13,200 people had attended a show in the 23.2 million dollar arena, the site of the 1976 Republican National Convention. No one was injured.

July, 1979/ G.M. PLANT ROOF COLLAPSES IN STORM

A windstorm was listed as the reason for the collapse of the 56,000 square foot General Motors roof near Kansas City. Due to the July 4th holiday, only several maintenence workers were in the building at the time.

July, 1979/ WOOD ARENA COLLAPSES, KILLS FIVE

A wood arch roof over a stadium at Rosemont, Illinois, collapsed and killed five workers. Damage estimates were set at nearly at 3 million dollars.

August, 1979/ ESTATE OF KENNEDY CENTER ARCHITECT TO PAY $25,000

A suit involving design deficiencies of the John F. Kennedy Center for the Performing Arts in Washington, D.C., and the estate of architect Edward Durell Stone, whose architectural firm designed it, was settled. The estate agreed to pay the government $25,000.

September, 1979/ NEW YORK ARENA BEING ANALYZED FOR SAFETY

Concerned over how much load the 1.5 acre space-frame can carry, owners of a 7,200 seat arena called in outside consultants to analyze the six year old structure. Cracks began appearing in concrete beams shortly after completion of the building.

October, 1979/ COLLAPSE OF MALL FRAME KILLS ONE

One worker was killed, another injured when part of a shopping mall under construction in Annapolis, Maryland, collapsed during steel erection.

February, 1980/ EL PASO CIVIC CENTER SUIT SETTLED

An out of court settlement was reached regarding the lamellar tearing of structural steel sections in the El Paso Civic Center.

February, 1980/ OVERPASS COLLAPSES, KILLS TWO

A cantilevered sidewalk portion 147 feet long fell from an Ohio overpass, killing two persons. The occupants of a passing car were killed as the automobile was cut in half by the falling concrete.

March, 1980/ KEMPER ARENA REOPENS WITH NEW ROOF STRUCTURE

After repairs costing 36 million dollars were made, the Kemper Arena in Kansas City was reopened.

March, 1980/ ARENA REPAIRS NEEDED, SAYS ENGINEER

The seven year old Veterans Memorial Arena in Binghamton, New York, required extensive repairs to bring it to workable shape, according to a consulting engineer's recommendations. He stated that a new 206 x 315 foot roof over the existing roof was needed since the original space frame roof was unsafe.

May, 1980/ MONTREAL STADIUM NEEDS ROOF REPAIRS

The Montreal Olympics Stadium required $72,000 in repairs to prevent a future roof collapse, according to published reports. The 16,000 seat stadium was unfinished.

May, 1980/ ATLANTIC LOADING DOCK ROOF COLLAPSES

The steel frame canopy over a loading dock at an airlines terminal under construction in Atlanta collapsed and injured 10 workers.

July, 1980/ CONGRESS HALL COLLAPSES IN WEST BERLIN

In a preliminary report, fluctuating stresses and corrosion to prestressing tendons were alleged to be the main cause for the collapse of a portion of West Berlin's Congress Hall. Inadequate protection against moisture contributed to the problem. One third of the 37,000 square feet concrete shell roof over the auditorium fell, killing one person and injuring five. Design errors were charged.

August, 1980/ BRIDGE COLLAPSES

Five days after workers poured a concrete deck, a 221 foot span of the 1,000 foot bridge collapsed, sending 620 tons of debris crashing into the American River 30 feet below.

August, 1980/ BAD CONCRETE DAMAGES 100 BUILDINGS

Several million dollars worth of damage to about 100 buildings in the San Francisco Bay Area was traced to discarded refraction bricks that were accidentally dumped into an aggregate pile at a California cement plant.

August, 1980/ SUITS BEGIN IN KEMPER ARENA COLLAPSE

It may be several years before all the information is gathered for Kansas City's suits against the architects, engineers and steel contractors. The Kemper Arena roof collapsed in June of 1979 year and lead to damage claims totalling 8.79 million dollars.

September, 1980/ SINKING BUILDING SUITS SETTLED

A $750,000 out of court settlement was accepted on the sinking $185,000,000 Federal Court House in downtown Philadelphia. Negligence was charged in the construction of the building's foundation.

September, 1980/ BEAM COLLAPSES, KILLS ONE WORKER

A beam collapsed in a warehouse under construction in Brattleboro, Vermont, killing one worker and injuring four others.

September, 1980/ STEEL FRAME COLLAPSES

Claims that a high gust of wind knocked down the one story 3.5 million dollar building came under investigation. Ten workers escaped injury, while several others were injured as they rode the steel frame down to the ground. Damage estimates were placed at $200,000.

September, 1980/ DEFECTIVE BEAMS SHUT DOWN CALIFORNIA SCHOOLS

Five Northern California school buildings with precast prestressed roof beams were shut down in a precautionary closures brought about by a statewide inspection. The inspection was undertaken following the collapse of an earlier high school gymnasium roof where no one was injured, since falling fragments an hour before the main collapse gave ample warning.

October, 1980/ DOMED CONCRETE FORM COLLAPSES

A large form for a concrete dome in Dennison, Iowa, collapsed, injuring three workers. The dome, 190 feet in diameter at its base was the largest of its kind.

November, 1980/ SIX-STORY BUILDING TO BE DEMOLISHED IN CALIFORNIA

After the first story structural columns collapsed during the Imperial Valley earthquake, the recently built structure was scheduled for demolition.

December, 1980/ ITALY QUAKE DAMAGE 24 BILLION DOLLARS

A major earthquake which struck southern Italy, measured between 6.5 and 6.8 on the Richter scale, left 3,000 dead, 1,500 injured, and 7,000 missing. Nearly 300,000 persons were left homeless.

December, 1980/ BUILDING COLLAPSES. STRUCTURAL DEFECT BLAMED

After shaking for four minutes, half of a 7-story wing of an Argentine air force building suddenly collapsed, apparently because of structural defects. 16 persons were killed, and 32 seriously injured.

December, 1980/ CIVIC CENTER STEEL STRUCTURE NEEDS REPAIRS

As construction limped along at the beleagured Worchester Civic Center in Massachusetts, a legal battle raged over the alleged presence of critical flaws in the original design and fabrication of the steel frame. The suit involved "more lawyers and engineers than there are bolts in the building," according to one source. The 15.5 million dollar project cost has soared to 18.5 million dollars. The city replaced all 159 steel joists, strengthened the trusses with 130 tons of steel, and rechecked all the bolts.

December, 1980/ FIELD HOUSE ROOF STRUCTURE NEEDS REPAIRS

A 16 million dollar Lake Placid field house required strengthening of structural trusses due to alleged faulty connections and flaws in the construction, according to investigators who reviewed the structure.

March, 1981/ BUILDING CLOSED. SEISMIC SAFETY QUESTIONED

A major California bank closed its distribution center in Northern California because engineers found that its prestressed precast roof could not withstand a strong earthquake.

April, 1981/ COLLAPSE OF FLORIDA CONDOMINIUM KILLS 11 WORKERS

Investigators from federal and local agencies searched the wreckage of a five-story condominium in Cocoa Beach, Florida, to determine the cause of the collapse. As crews were pouring concrete on the roof, the entire building collapsed, stacking floor slabs, one on top of the other. The project cost was estimated at 5.5 million dollars.

May, 1981/ GARAGE FLOOR FALLS IN CHICAGO

A 20 x 29 foot portion of a Chicago underground parking garage collapsed. No one was injured.

June, 1981/ SINKING STADIUM NEEDS REPAIR JOB

The stadium at Arizona State University, originally costing 4.5 million dollars, required a 3.5 million dollar repair job due to settlement.

June, 1981/ BUILDING SETTLES 4 INCHES

In Massachusetts, settlement in a recently completed government building caused cracks in the foundations. An investigation was begun.

July, 1981/ KANSAS CITY HOTEL WALKWAYS COLLAPSE

The collapse of walkways at the Hyatt Hotel in Kansas City killed 111 persons and injured scores of others. It appeared that connections at the supporting steel rods failed.

October, 1981/ CHURCH FLOOR COLLAPSES IN NEBRASKA

While church members were enjoying a turkey dinner at a church in Ainsworth, Nebraska, the floor collapsed, falling into the basement. Shortly before the failure, children had been playing in the basement. The building was one year old.

October, 1981/ POSSIBLE DESIGN ERRORS IN NUCLEAR POWER PLANT

Utility company engineers discovered what could be "potentially critical" errors in the design of structural reinforcements intended to ensure that a California nuclear power plant could withstand earthquakes, according to the Nuclear Regulatory Commission.

October, 1981/ FORM COLLAPSES, KILLS TWO

Two men were drowned in concrete and ten others were injured when 25 cubic yards of fresh concrete were dumped upon them as a steel form collapsed at the construction site of a small hydroelectric plant addition to the Greenup Lock and Dam on the Ohio River.

December, 1981/ CHICAGO HYATT REGENCY CLOSED FOR MAJOR REPAIR

A consultant for the City of Chicago warned that a structural collapse was "imminent" at the Chicago Hyatt Regency Hotel cantilevered bar, and could have reached the magnitude of the Hyatt walkway collapse in Kansas City. The city ordered the hotel to close off and shore up a suspended area it deemed dangerous until it could be strengthened.

January, 1982/ SIX INJURED IN CONDO PROJECT

A freshly poured section of concrete collapsed at a Los Angeles construction project, injuring six workers with a tangle of scaffolding, plywood and decking.

January, 1982/ CONCRETE PADS FAIL IN MIAMI PROJECT

The contractor was asked to replace 72 faulty concrete beam bearing pads. Earlier, several concrete beams had cracked and had required structural repairs.

January, 1982/ CRANE COLLAPSES, KILLS THREE

A construction crane at a 42 unit condominium site in Vero Beach, Florida collapsed and killed three workers and injured three others.

January, 1982/ COLORADO STORM DAMAGES BUILDINGS

Government estimates revealed that more than 50 houses were severely damaged and forty percent of the buildings in Boulder, Colorado were damaged during a violent windstorm.

April, 1982/ CHICAGO HYATT REPAIRS EXTENSIVE

Repairs to the suspended cocktail lounge in the Hyatt Regency Hotel were found to be more extensive than originally anticipated. Officials explained that truss members and joints, as well as the primary tension members required additional strengthening. An outdoor walkway was also required to be repaired.

June, 1982/ CHICAGO HOLIDAY INN GARAGE SETTLES

Footing failures were believed to have caused a six inch settlement in the downtown Chicago garage. Driven piles were placed to rescue the structure.

Why do we accept this shoddy performance of our buildings? Why does it seem that there is never time enough to do a project correctly the first time, but always time to pick up the rubble afterwards and rebuild? Would we accept this type of performance from other areas of our society that concern our life safety? We have taken giant steps in our transportation industry to improve safety. Airline accidents are very few. Automobile safety has gained much attention in recent years and has been dramatically improved. Safety concerning radiation from our nuclear power plants is constantly in the headlines. Industrial accidents in manufacturing plants are always of high concern, and great progress has been made. It is now the time to turn some attention and effort to a long neglected industry, and begin improving the stability and safety of our future buildings.

Photograph on preceeding page:

January, 1978. The 2½ acre space frame roof of the Hartford Civic Center Coliseum collapsed. The main center portion of the roof fell 83 feet to the 12,000 seat arena below. There were no injuries since the arena was not in use.

—Courtesy of The Hartford Courant.
Photographer: Tony Bacewicz.

Chapter Two

WHAT THE PUBLIC HAS A RIGHT TO EXPECT

No one can disagree with the statement that buildings should be designed and constructed to protect the safety of all users and occupants. There is absolutely no excuse for any roofs to fall, any walls or columns to cave in, or any concrete panels to fall from buildings to the sidewalks below, unless of course the building is subjected to an extreme natural disaster.

Every building failure with which I am familiar could have been prevented with relative ease. With few exceptions, each and every failure can be traced to insufficient time spent in designing or in checking the plans and details, or to the lack of proper supervision during construction. Only in very few cases, have roofs fallen because the snow depth was greater than recommended by the building codes. Even these cases it is an unacceptable excuse, for the following reasons.

It should be understood that building code requirements are the absolute *minimum* loads that a building is required by law to support. The fact that a building was designed per building code requirements does not at all mean that it is safe. After all, the great majority of the buildings which have failed did, in fact, comply with these requirements. They all had "building permits".

The truth is that for an extremely small additional cost, the load carrying capacity of any roof span can be increased manyfold — enough to support any conceivable snow load that may occur but once every 50 to 100 years. This additional roof capacity should be mandatory, especially in the design of large span-roof structures that may loom precariously over thousands of people in coliseums, auditoriums, and arenas.

When all the cost factors of a building are considered, such as the land, heating and air conditioning, lighting, finishing materials, furnishings, landscaping, parking lots, plumbing, sewers, insurance, and financing, one soon recognizes that the actual cost of the structural roof is a very small percentage of the total cost — perhaps as little as five percent. If this one portion of the building would be strengthened to carry three times the minimum snow load, the overall cost of many buildings would not increase by more than two or three percent! This is considerably less than the current annual inflation rate. (When replacement cost of a collapsed roof is considered, the overall cost is actually greatly reduced). In any case, would it not be worth it for all concerned to have the peace of mind that the roof over theirs heads remains where it is intended?

Again, the major problem seems to be a common reluctance of owners and developers to spend any more money than is absolutely necessary on those portions of the building which cannot be seen or felt. I have personally been involved in complete redesigns of buildings to save a mere five cents per square foot. A common occurrance is for fabricators of structural roofs to bid for supply of roof systems, providing minimum designs of roof members that barely meet requirements to pass the municipal building codes and thus secure a permit. Beware of unexpected pigeons that might land on such roofs!

Similar situations have taken place with the concrete flat slab floors in many buildings. In the original plans the structural engineer may well have used somewhat more reinforcing steel than required by the building code to reduce floor deflections. Once again, the bidding concrete contractors who hire their own engineers, redesign the floor specifications and reduce the reinforcing steel. These re-calculations fatten the pocketbooks of the contractor, as well as the owner. After several years, when long term deflections have taken place, many of these buildings are burdened with sagging floors, tilted walls, and windows and doors that do not operate properly. Obviously, these design revisions do not even consider potential future problems. What false economy! (But what nice added profits for the builders and owners!)

Several of the structural failures mentioned earlier concerned heavy concrete or marble wall panels that fell from high-rise buildings to the sidewalks and streets below. Fortunately, in the cases cited no one was injured, but what a catastrophy this could be for anyone walking along the sidewalks below. Certainly, it is not that difficult to design and build panel anchorages that will never come loose.

Some of the other failures mentioned dealt with very long span space frames. This is truly the most dangerous failure of all, due to the fact that many people may occupy the building at any given time. The dreadful potential for thousands of injuries or deaths is always present. What makes the situation an even higher risk is that the long span building type has a relatively short history. The designers are always looking for new economical design methods, some of which are quite complex in theory. Structures of this magnitude require rigorous analyses, and no expense should be spared in guaranteeing that every detail is carefully considered. After completing the design, the work should be checked and then rechecked. If the shape is unusually sophisticated, models should be built and tested. Of course all of this takes time and money. Unfortunately, even on large projects most of the work is rushed, and necessary funds required to cover any additional design costs are rarely provided.

There have been so many failures in long span structures that some news journalists have questioned our technological ability to safely build such edifices. Even the ability of engineers to competently design these buildings has been challenged. Nothing could be further from the truth.

The owners of these enormous buildings, whether they be a city, large corporation or powerful individual, are all accustomed to calling the shots. This makes sense to many people since these influential bodies have both the money and the entrusted power. They are thus able to intimidate architects and structural engineers into meeting their timetables and budgets, regardless of how unrealistic they might be. The time and fees required to do a proper and consciencious job are rarely made available. Rushed deadlines are often prearranged before the engineer's involvement with the project, and these deadlines become the single most important item from that point on. A standard fee is offered on the basis of the general building type, with little regard for the complexity of the specific project.

As work progresses on the design, a continuing struggle occurs among the unrealistic schedule, the low fee, and the consultant's attempts to keep the design costs in check. After all, the engineer is in business, and if his firm does not make a profit on the job he could find himself in great financial difficulty. Because of the tight and restrictive fee, the engineer is pressured to cut corners wherever possible. The fee is usually fixed, and automatically determined by the number of manhours spent by the design team on the specific project before going into the red.

Unfortunately, the same problems occur during the construction phase. Poor inspection and field supervision is quite common. Often, important changes are made by the contractor during construction which are not cleared by the engineer.

As the previous listing reveals, most building failures actually take place during construction. Complete and detailed supervision, continuous, if necessary, should take place by the structural engineer or representative. This is the only practical way to assure that a project is built exactly as detailed on the engineer's drawings. At least, all changes should have a full written approval of the engineer, and the only after careful analysis.

People should have a right to attend gatherings in large span structures without the anxiety of a roof collapsing upon them. They should have a right to walk down streets without the worry of glass or conrete panels plummeting from a high-rise building. They should also have a right to expect that floors will not sag or be dangerously bouncy. Likewise, workers at construction sites should have the right of working in relative safety. But, unless major changes are made in our system, there is very little hope that the present conditions will improve. We may expect to continue reading of and witnessing building disasters directly caused by decisions of those persons involved in the building industry who are least knowledgeable of building safety.

Photograph on preceeding page:

October, 1981. While church members were enjoying a turkey dinner at a church in Ainsworth, Nebraska, the floor collapsed falling into the basement. Shortly before the failure, children had been playing in the basement. The building was one year old.

—Courtesy of the Ainsworth Star-Journal

Chapter Three

SO YOU WANT TO BUILD A BUILDING!

In almost every direction we look, we can see new construction in progress. Those of us not involved with the building industry are often completely uninformed on the procedures used and the ways in which all of this comes together. Who, for example, actually designs the buildings? How are the "blueprints" drawn? What does the architect actually do? What is a structural engineer? How do the contractors put all the pieces together?

Surprisingly, many people think that an architect is the person who draws blueprints. Although this view is partially correct, it is far from being complete. Initially, the architect and client discuss the requirements of a project — the costs, client preferences in style, and the functions of the completed building. The architect then proceeds to prepare a series of schematic drawings, showing a set of proposed solutions to the client's needs. These are usually very visual and artistic renderings, sometimes even employing scale models that are prepared to allow the client to better envision the building.

Once the client approves these preliminary plans, the architect begins preparing a set of architectural "working drawings". At this stage, unless the architect's firm is large and has in-house engineers, the architect engages the services of outside consulting engineers. A structural engineer then completes the structure's design, while mechanical and electrical engineers design the appropriate air conditioning and lighting while other specialized experts are called upon as needed. Each of these independent consulting engineers then prepares their own drawings, in continuous coordination with the original architectural drawings. They each prepare their individual engineering calculations as required. For example, the structural engineer will apply knowledge of engineering to determine the actual sizes of each and every structural member and its connection throughout the building. The project engineer also determines the loadings on the beams, columns, and footings, and selects the sizes and strengths of all structural members, whether the materials are wood, concrete, or steel. The engineer also determines the number and size of bolts or weld sizes; provides bracing details for wind and earthquake loadings, and calculates the amount of reinforcing steel required for the concrete elements.

After the working drawings are completed and coordinated, along with a set of typed specifications, the project is submitted to the local building department for its review and approval. Only after all the plans are approved can any project be issued a Building Permit, afterwhich the plans are printed and distributed to contractors to obtain firm bids for its construction. When the successful bidder receives the contract, construction work begins.

Periodic field inspections of the construction site should be made by the architect to see that the plans are being followed. During construction, he usually approves the owner's payments to the contractor at completion of various phases of the work.

This is a proper place to briefly mention the educational requirements for architects and structural engineers. Their pre-college courses usually include history, languages, social studies, physics, mechanical drafting and abundant courses in mathematics. Architects and engineers generally take a four to six year study program at an accredited architectural or engineering college. Thier courses include six basic areas: 1. architectural history, 2. architectural design, 3. structutal analysis, 4. building equipment, 5. visual design, and 6. city planning. The structural engineering student will normally take a course in civil engineering, with specialization in structures. Some colleges offer a degree in Architectural Engineering. In order to open an office, the architect and engineer must be licensed in the respective field in the state of the practice. Although the laws vary somewhat from state to state, most states require four to five years of professional training after completion of high school, and three to four years of approved qualified experience. At that time, one may be eligible to take the professional registration examination, which can last from one to four days, depending upon the state involved. In California, for example, a licensed structural engineer must first meet all the requirements and then pass the examinations to become a licensed civil engineer. Then, after three additional years of experience are required before one is eligible to take the two-day examination to become a registered structural engineer.

The overwhelming majority of architectural and structural engineering consulting firms are small operations employing usually less than ten persons. Only a handful of firms throughout the country can be classified as any larger. In this profession, a "large" firm may consist of 30 to 100 employees. This is quite a contrast to the enormous aerospace industry, some of the huge defense-oriented companies, and the other large engineering firms that design oil refineries and nuclear power plants. Some of these corporations employ thousands of engineers and technicians.

One of the astonishing facts about the building design profession concerns the unusually low salaries paid quite uniformly to all members of the design team. One must first understand the unique responsibilities shared by these designers in creating and detailing plans for buildings which are intended to endure for more than one hundred years. During this period, thousands of occupants may enjoy the benefits of these structures. However, the average construction laborer, who works for the contractor earns substantially more in hourly wages than does the typical graduate architect or engineer who works for the professional design firm. To make matters worse, the wage spread is even greater when compared with considerably greater earnings of other so-called "professional" trade workers, such as carpenters, electricians, and plumbers.

In order to gain a better understanding of the problems faced by each of the groups involved in the design and construction of a building, let us briefly analyze what is generally in the mind of each. Of course this may be fanciful conjecture, but nonetheless probably accurate in most instances.

The developer or owner of the project will prefer to build the most with as little "up front" outlay of money as possible. Usually the owner is dependent upon large financing from various lending institutions for construction funds as well as for the longer term operation of the structure. The initial investment is often limited, and since the major portion of these investments covers architectural and engineering fees before construction begins, the owner faces pressures to keep these expenses as low as possible. This, of course, leads to indiscriminate fee shopping in many cases. While design fees represent the largest percentage of the initial outlay of funds, these fees are, in fact, very small in proportion to the overall cost of the project. One way to arrange an adequate fee for the design professional would be for half or so of the fees to be payable up front, with the balance due immediately upon funding of the construction loan — possibly through a some type of escrow account. Provisions could also be made for some of this balance to be paid in the event the project is not built. (The portion of the fee allocated to construction supervision and "responsibility" would, of course, be withheld.) More sophisticated developers have now come to realize the importance of selecting qualified architects and engineers for their projects, and not just someone who will work for a low fee. For their own sake, owners should always keep a close watch throughout the design phase, reviewing tentative costs of construction, so no unpleasant surprises appear when the final construction bids come in.

The architect for the project basically wants an attractive, functional building which will meet the client's needs. Most architects would like to have a sense of pride in the finished building, with the hope of obtaining future commissions from the same client as well as referral clients. The architect is also interested in the long-term endurance of the building, and of course does not want any part of consultant created problems. The architect realizes that it is an advantage to engage the services of competent engineers to avoid possible future troubles, but also realizes a concern about making a fair profit. Like everyone, the architect does not want to pay more than is necessary; since the consultant fees have to come out of the overall fee from the client. Many architects are well aware that if their own fees could be increased, they would then be able to raise their payments to the engineer consultants and create a building of higher quality. At the present time, architects are not completely comfortable with the work of consultants, including the structural engineer, but considering the circumstances of their low fees and the rushed schedules many realize that this is probably the best that can be expected. For the same reason, architects are probably not always very satisfied with the work of their own employees.

The structural engineer for a project would usually be happy just to know that the building will not fall down. They probably have their share of nightmares about buildings collapsing, perhaps because of a draftsman forgetting to place enough bolts in one of the connections. (I know that I have had my share of nightmares). Engineers usually breath a sigh of relief whenever one of the buildings they have designed has been completed without complaints about sagging roofs, bouncy floors, or cracks in the concrete. In general, the thoughts of structural engineers coincide with the architect's; they would like to have some profit left over at the end of the job, and they would like to be perfectly content to go through life without ever having to see another lawyer to defend against lawsuits, justified or not.

Above everything else, engineers probably would like fees large enough along with adequate time to do a throughly professional and complete job on every project in order to avoid future problems not only for their own professional protection, but for that of all concerned.

The contractor is hired by the owner or developer to construct the building in keeping with the plans, details, and specifications set forth by the architect and engineer, and is typically selected on the basis of the lowest bid. The contractor should have a background of experience with similar type and size projects before being chosen to present the bid, and should also have the financial responsibility and ability to assure the completion of the work. The basic reason that the selection of the contractor is often determined by the lowest bid is that each portion of the building project is specified right down to the last nut and bolt. The contractor knows exactly what must be provided, and is therefore able to determine an exact price for the work. In contrast, the reason why this kind of bidding method is not applicable in selecting an architect or engineer is that the exact details and scope of the design work can rarely be spelled out. There is a vast difference in what one design professional may produce compared to another. The procedure and amount of detailing may vary, as well as the overall interpretation of the methods available to meet the client's needs. Of course it is possible for the architect and engineer to "shortcut" a job by not providing all of the details necessary for absolute clarity. (In the chapter concerning fees, I have facetiously illustrated some examples of these shortcuts.)

The next time that you see a building under construction stand there for a while and watch. Realize that every single part that goes into the building has been designed and detailed by the architect and engineer to every last bolt, nail and screw. The structural engineer has determined the exact number of bolts to be used in each connection supporting a steel or wooden beam. Every reinforcing bar has been detailed in size and location. Each beam and slab has been carefully thought out with regard to its load carrying capacity. Count the number of different parts used throughout the building, and you will get an idea of the enormous complexity of the structure. What you don't see are the reams of technical calculations that were necessary to justify the strength of each item. An error in any one of the small parts could be responsible for a structural failure. It's something to think about.

Photograph on preceeding page:

June, 1979. The 17,600 seat Kemper Arena in Kansas City was the site of a major collapse when the roof structure crashed 95 feet to the floor below during a rainstorm. Two nights earlier, 13,200 people had attended a show in the 23.2 million dollar arena, the site of the 1976 Republican National Convention. Suits, claim damages and lost revenues totaled 8.79 million dollars.

—Courtesy of The Kansas City Times.
Photographer: John Spink, © 1979.

Chapter Four

"We need the plans by the 15th. If you can't meet my Schedule, I'll find someone who can!"

How important is a major building which is to be used by thousands of people over a period of 50 to 100 years or more? Does it deserve sufficient design time to assure a complete, thorough job? Unfortunately, many major buildings often have elements that are hastily designed and constructed.

The general attitude seems to be one of just getting the construction overwith as quickly as possible and then cover it up. Nothing has more priority than completion schedules — especially when interest rates are high. The builder does not want to pay any more financing charges than absolutely necessary on his construction loan. Many construction superintendents have lost their jobs by trying to do too thorough a job, and thereby not meeting the schedule.

I am not judging other engineering or architectural offices by my own shortcomings during the times when I was rushed into impossible schedules. I have worked with many owner-developers through the years, and this driving pressure is very common. During a period of my engineering career, I worked for the California State Office of Architecture and Construction. There I reviewed and checked plans for public schools, submitted by many of the best architectural and structural engineering offices in the country. Over a six year period, I had analyzed plans for hundreds of school buildings, and not once did I find a set of plans and details without either glaring errors or minor to moderate misjudgements. And it must be kept in mind that the design of schools in California is done by the most competent firms, with the least pressure to hurry the job. The school plans done by these firms are far superior to the plans usually turned out by engineering offices for commercial buildings.

When the boss speaks, everyone listens. In this case, the "boss" is the builder or owner-developer. He is the one who pays the bills, and consequently, is the one who calls the shots. Owners often pressure the architect into a schedule which is based upon everything going perfectly along the way. If there were no delays or changes, it might just be possible to come close to this schedule. However, I have seldom worked on a project without some unforeseen problems or changes which cause delays.

The very first question that is asked of the structural engineer after the architect has agreed to a time schedule with the owner of the project is whether or not the engineer can commit to this schedule. It is a simple question. If the reply is "no", then someone else is called in. If the engineer is low on work at the time, he will have a hard time in not committing to a project schedule. He may plead for more time, but more often than not, he will agree to the schedule while keeping his fingers crossed and hoping for a time extension as the job progresses. He knows, better than anyone, that if he wants to work in this field, he had better meet the schedules, or else return to school to study medicine or law. (Or perhaps work in real estate sales.)

When I operated my office, I was unable to hire or keep good competent engineer employees if they could not work at a super speed. It was like trying to only hire a secretary who could type over 100 words per minute. The engineer had to know the right stuff. He had to have a thorough background in the field, and, of course, good references. Above competency, he also had to be able to "jiggle" out a quick engineering calculation or a speedy little structural detail. He might have been sufficiently qualified to singlehandedly design and draw up the World Trade Center, but if I would have had to lose money because an engineer was too slow and thorough, I could not afford to employ him. Let me briefly explain what "losing money" means in this case. I've done it many times. When I charge a fee of $10,000 to do the structural engineering for a project, and I pay out $15,000 in salaries alone to get the job done, and then I have to wait two years to collect my last payment on the $10,000 — that is what I mean by losing money!

Because of the rushed schedules on so many projects, it is my opinion that the engineer and his employees cannot help but become careless. They are so pressured for time that they begin to crank out engineering details at an assembly-line pace. All sense of responsibility becomes temporarily diminished to the point where good engineering judgement becomes secondary to speed. In some instances, only a few minutes may be allocated to each detail. If the detail is incorrectly designed, a failure could occur in that portion of the building. It takes a failure on only one bolt or weld to bring down a roof. It has happened many times before. However, the time schedule becomes omnipotent, and the seeds for a possible tragedy in the future are planted.

Years ago, when I was employed as a "semi-junior" structural designer, I was asked to design and detail a warehouse with an attached office building wing. Because of the usual hurried schedule, I quickly drew up all the precast concrete wall panel details for the front exterior wall of the office building. Within several days, construction workers were busy at the job site pouring the concrete into the flat wall forms. Several days later, the concrete had hardened, and the wall panels were tilted up into final position. That same day, I received a panic telephone call from the construction superintendent. "How in the hell are we supposed to get into the damn building?" he wanted to know. "Climb in through a window?" In my haste, I had forgotten to put the front door opening on my wall panel details! No one had taken the time to check my plans before pouring the concrete, and the entire front wall of the office building was built without a front door! The contractor had to jack-hammer an opening in the concrete. As unlikely as it may seem, this actually happened!

Chapter Five

"Your fee is too high!"

Presently, all design fees are paid to the architect, who in turn pays consultants such as the structural engineer, the mechanical and electrical engineer, and others. As a general rule these fees amount to between 2 and 8 percent of the bare building cost. This cost is effectively reduced to between 1.5 and 6 percent of the project cost when the cost of the land is added in. This percentage is even further reduced when finishing costs such as furnishings, partitionings and carpetting are included.

The amount allocated to the structural engineering firm for an entire structural design of any given building is usually about one-eighth of the total fee. In other words, a building project worth $1,000,000, includes a fee of $1,500 to $5,500 for the structural engineer. For the work, an engineer may receive a final net profit of between $250 and $1,000 after paying salaries and overhead expenses. Keep in mind that most structural engineering offices consist of less than 5 employees. By way of comparison, the real estate company which sells the project will normally realize a fee of about $60,000, and the sales agent who lists and sells the project may receive about $30,000!

To gain a better perspective on what the architect's "normal" fee is for a typical project, consider the following: one of my architect friends once remarked that his fee was approximately equal to the sewer hookup connection cost for the building! This made me stop and think how I could equate my structural engineering fee on the building. After some lengthy calculations, I found that my fee was equal to the cost of one lonely steel beam in a $750,000 project! I drove out to the job, and stood there and looked at that beam... and as I looked... I reflected upon the many days and weeks of engineering effort that had gone into the project. I thought about the calculations; the meetings with the architect and the client; the problems with the building department and the stormy scheduling problems. I looked at the beam again... then turned and walked away.

If the structural engineer's fee was even tripled, the net effect on the project cost would be barely discernible, yet this increase would result in accomplishing a more complete, safe, and thorough engineering job. It would have the additional effect of attracting more qualified individuals back to the profession; especially those fine engineers who left the building industry to work for the mammoth aerospace and defense firms for higher salaries.

One may ask what assurances there would be that higher fees would automatically result in higher quality work. There are ways to insure this, which I will discuss later. One thing is certain, however, low fees will absolutely guarantee that most of the work done by engineers will be inadequate and improperly checked.

It cannot be overstressed that the salaries in this field are very low — lower than the wages of construction workers. It is a discomforting fact for the designer to realize that all of the educational background; plus additional studies to pass architectural and engineering licensing examinations, and years of experience in creating new designs with new technologies, all add up to make it unlikely to earn a fair wage. The typical engineer earns far less than the construction worker who hammers in the nails, tightens the bolts, or plasters the wall. How can the ambitious architect or engineer employee cope with a minimum standard of living and rationalize the situation? One common way, rampant within the industry, is "moonlighting." This side work can be done during evenings and weekends on private projects or for other architects or structural engineers.

It is really difficult for the non-building professional to fully comprehend the vast amount of building plans produced annually by people who work part-time in order to supplement their basic full-time salary. Many millions of dollars worth of construction is executed every month from plans designed and drawn by moonlighters. It is true that these projects may not be major high rise structures or enormous sports stadiums, but some of these buildings are substantial in size.

I would venture to say that most employees in the field do some moonlighting either regularly or as the opportunity arises. There are basically two reasons for this. First, and most apparent, they need the extra income. It is frustrating to have spent so many years studying for a profession that you believed in, only to discover, later on, that you are but one step above mediocrity — that you may never be able to afford to live in any of the nice custom-built homes that you have designed. After several years of struggling in the field, many architects and engineers realize the inherent limitations of their work in the building industry. They often feel helpless and trapped; realizing that if they are to reap the niceties of life they must take on some moonlighting work. The second reason is that architects and engineers often hope to open their own consultant or design office some day, and believe that if they begin to attract clients on a part-time basis, it might be easier to make the future transition.

Unfortunately, this flood of moonlighting work within the building industry has added to the problems of the professional full-time office. A client that hires a part-time architect or engineer does so knowing that the work will be done at a much lower fee. Why else would an owner not use the services of a full time professional? This, in turn, creates additional competition for the full-time office, and causes a downward pressure on fees which are already below par. An even worse and dangerous consequence of this cycle is that part-time workers are likely to turn out inferior work. They are often less skilled and work without experienced supervision. In this situation the plans of the part-timer will definitely go unchecked. In the end, the client saves several dollars in fees, and could care less; while the moonlighter is able to have the supplementary income not available from a full-time job.

Imagine this moonlighting situation happening in other professions. Would one go to a part-time lawyer to save a few dollars? Of course not. We all recognize that in the long run we would not save any money. Clients are usually able to see for themselves that the wise decision is to pay for the services of a top-notch legal advisor. A part-timer is unthinkable. But in the construction business, if the plans are poorly executed, yet passable for the builder, what might happen in the distant future becomes a meaningless issue. The builder will most often sell the building anyway, and possible sagging floors and roofs will become someone else's problems. If a more serious problem arises, it will be blamed on the architect or structural engineer anyway. So why not save the money in extra fees? This is an enormous responsibility to put on a part-time "professional." In reality, the moonlighter rarely understands the full liabilities incurred, but more of this later. In any event, it is the public that loses, since the damaged victims of a building failure are rarely the same people that pay the architectural or structural engineering fees. They are usually innocent third parties that have nothing whatever to do with these fee decisions.

Fortunately, the type of person who enters the architectural and engineering field usually has the dedication and professionalism to do the best job possible despite the financial and time restrictions. Employees in this field might improve their condition by forming their own union. This would gradually drive up incomes, with even the possible side effect of increasing the income of the principals in these firms. After all, if owners would have to pay higher salaries, they would then demand higher fees to ensure an fair profit. Another outcome of unionization would be less moonlighting with all of its negative consequences. Unfortunately, employees in this field, as a rule, do not consider unions "professional," and since so many of them have high hopes of eventually opening their own offices, they often consider themselves as future employers, and as such, are frightened by the thought of unions "taking over" their profession.

•••

A common occurrence that happens during the course of any building construction involves the omissions or errors made on the original design drawings that are discovered during the time of executing the plans. These mistakes includes things like dimensions that are incorrectly coordinated between the various trades, mislabeled and ommitted beams, or even smaller mistakes like an arrow leading the builder to the wrong portion of a detail drawing. These kind of mistakes are common in the best of plans, and again, in my experience, I have yet to see a set of plans that was completely perfect in every respect.

In actual practice when an error is discovered, the contractor usually makes a quick determination of whether any additional expense or time will be necessary to make the correction. If there is an expense, the cost of correction is given to the owner-developer for approval. Upon receipt of the bill, the owner-developer may forward it directly to the architect-engineer for payment, or else, the owner may deduct the amount from the remaining unpaid fee, claiming that the extra costs were due to the engineer's error. Quite often, the cost of these items can exceed the net profit of the architect-engineer and in some cases the entire fee. The result is often a major loss for the building designer who may have devoted one or two years of time to the project.

Now let us analyze what happens in this situation. The owner-developer paid a low fee to the architect and engineer to begin with. In fact, more often than not, a large portion of the fee remains unpaid. The fee may possibly have been too low to provide for a complete, thorough service. There may have been insufficient time to do a first class job. Most likely, the plans were not checked, or if they were, the review was very minimal. In other words, payment was really made for a minimum type of service. Now that something goes wrong, the owner-developer fully expects the underpaid architect or structural engineer to cover the cost of correcting the problem. In these events the owner-developer may often obtain many extra items (if they were omitted) at no additional cost. In the operation of my own business I have paid plenty for such items. In one instance, the owner insisted that I provide the structural steel angles and bracings for a sign on a building which I engineered. He claimed that my drawings should have indicated the steel, and since they did not, he refused to pay the "extra." In effect, he received all the benefits of the materials and labor at no cost to himself. The building was small, and the cost of the extra steel far exceeded my firm's profit after several weeks of design work. On a larger project, this situation could drive the architect and engineer out of business and into bankruptcy.

An outsider who looks at this chain of events might readily say that the architect and structural engineer should not have taken the job to begin with if they could not perform the complete service. Unfortunately, this is the way the industry operates. This is a fact of life, and no one architect or engineer can change the system. If a sufficiently large fee is demanded, enough to really do a complete and 100% thorough job, only a cloud of dust would remain in the room as the potential client scampers off to the next architect and engineer down the street. As a bright philosopher once remarked, "if you want to stay in this business, this is how you play the game." Well, many excellent architects and engineers are continually leaving the business, unable to play the game by these rules.

It is all competition, isn't it? Let me say that there is nothing wrong with competition. After all, that is the American way. But then, would the public accept unregulated competition in the designing and building of our jumbo jet airliners? Would we accept a free-for-all of low bidding without some type of controlling agency to assure complete public safety? I doubt it.

We are all aware of the dangerous situation that might exist if competitive bidding was the determining factor in aircraft designing. Just think, the aerospace engineer with the lowest bid would win the contract to design a plane as he wished, using whatever shortcuts he could get away with, as long as the plane could be built from the plans. This is basically the situation we have with building design! What if there were no Federal Aviation Administration to review and approve every minute detail of aircraft designs? We would probably have planes crashing all around us (and the railroads would again be booming).

The reason that plane travel is generally safe is basically due to airplane designers who are are given enough time and the money to do a thorough job. All of their details are checked, reviewed, and checked again. Enforcement by the F.A.A. is complete in every detail — nothing is left to chance. Anything less than perfection is unacceptable. What a contrast to our present way of designing and constructing buildings!

Inertia is a strange thing. The dictionary defines it as an "indisposition to change." The status quo has been accepted for so many years by architects, engineers and owner-developers, that no one will seriously consider change. New engineers, eagerly opening their offices, follow in the footsteps of their predecessors. The firms that had previously employed them traditionally charged low fees and provided a minimum service. Now they, striking out on their own, do the same. People's actions are so predictable that they often operate within the established traditions of their trade with hardly a thought that their acts might not be appropriate for the technology or social climate. It brings to mind the story of a young housewife who always cut off the end of the ham before baking it. "Why do you do that, honey?" asked her husband. "Because my mother always did it," she replied. "Let's ask your mother." said he. "Why do you always cut the end off the ham before you bake it?" they asked her mother. "Because my mother always did it" said she. So all of them approached the grandmother. "Why did you always cut the end off the ham?" Grandmother replied, —"Because my roasting pan was too short!"

Rodney Dangerfield, a popular comic, made a career using the "no respect" syndrome. Many architects and structural engineers can honestly complain about the same feelings, only without the humor. They feel that the clients and potential clients have a low opinion of their professional responsibilities. The clients, on the other hand, do not understand how the design professionals can ask so much money for what appears to be so little work.

Much of the problem is due to an absence of understanding of the enormous amount of technical background, knowledge and experience involved. It is the responsibility of the architect and engineer to raise the public's awareness to these matters — perhaps even through commercial advertising. Public acceptance of the present day fee structure of doctors and attorneys is well established. But all too often, the fees paid to design professionals are regarded as necessary but unwanted expenses encountered along the way of a building project. Often, all that is really desired is sufficient information on a set of plans for procuring a "building permit."

For example, I once rendered a fee proposal of $3,000 for the structural engineering design on a particular project, worrying all along that the proposed fee might be too low to cover my minimum basic expenses. When presenting my potential client with this fee, I was amazed to discover that he had expected only to pay a maximum of $350! He could not begin to understand the $3,000 fee. To this day I have little doubt that if he would have looked long enough, he probably could have found some moonlighting engineer who would give him a minimum amount of work for the $350 fee.

One of the things that encourages a client to fee-shop to such a large degree is the great discrepancy in the rate structure of the various offices (and moonlighters). When I first opened my office, I must admit that I did not know what a "proper" fee would be for a particular job. I called around and talked to other small engineering offices to see what they would charge for a similar type of job, and was astonished at the different answers I received. The fees varied greatly. The one that most impressed me at the time was the suggestion that I should calculate how much time the job would take, and then multiply my hourly rate by the estimated number of hours. One project came into my office, and I gave a fee proposal of $600. I was given the job. Later, I found out that two other engineers had been asked for proposals. One quote was for $1,200 and the other was for $2,400. Is it any wonder that clients are motivated to shop around for the lowest fees?

The following dialogue is a typical discussion that takes place between the design professional, architect on structural engineer, (Design pro.) and a prospective builder, (Client) in negotiating the fee:

Design Pro

"My fee would be $15,000 for this project."

Client

"Your fee is WHAT? This building is just a simple box, you know. The most I can pay is $9,000."

Design Pro

"I just can't do it for that."

Client

"You know, I've got a lot more work coming up that I'd like to use you on."

Design Pro

"Well, I guess there really is a lot of repetition in the job. Alright then....$11,000."

Client

"Look now, don't give me anything fancy. Just give me the minimum I can build the building with. Just some plans and a cross section. The last architect and structural engineer I used gave me so many details, my contractor didn't even use half of them!"

Design Pro

"Was that the building on Elm Street with the sag in the front canopy that caused the front windows to pop out?"

Client

"Yeah, that's the building. But you know, I sold the building before it started to sag, so I came out smelling like a rose!"

Design Pro

"Yes, I heard that the architect and structural engineer got stuck for the repair bill."

Client

"Look... you want this job or don't you? $9,000."

Design Pro

"O.K.... $9,000."

Client

"By the way, I need the plans by the 10th. I'll be able to pay you if... ah... I mean ***when*** *I get the construction loan."*

Now don't laugh. This type of negotiation takes place daily throughout the country. To further illustrate how ludicrous it is when a developer or owner is shopping around for the lowest fees, I would like to present the following list of items that can be used by the design professionals in their preparation of plans. Keep in mind that the plans as described will in every way meet the requirements of the applicable building code, and still be used for construction of the building. Beyond those two items, I can make no further promises!

1. Structural engineer to design the largest, most heavily loaded beam, and then make all the rest of the beams the same size. This will save a lot of time! **Who cares if it costs a little more to build?**

2. Structural engineer to follow the same procedure as item one above with respect to columns, footings, joists, rafters, and other structural members. More time saved! **Besides, what does the client expect for such a low fee?**

3. Architect to make entire roof flat except for drainage, and do not use any roof overhangs or canopies. This will keep the drawings simple, especially if the building is square! **Anyway, looks aren't everything!**

4. Architect to make all windows and doors same size for simplicity. Also, use no balconies or other similar devices **which only tends to complicate the drawings.**

5. Structural engineer to draw all plans on a very small scale. This will save tracing paper, blueprint costs, and will help to cover up small errors in plans, since some of the notes will not be very legible. **When in doubt about any item on the plans, a small finger smudge by the draftsman will usually hide any potential problems.**

6. Architect to give only the overall outside building dimensions on plans, with a note for contractor to adjust everything in the field to fit. This will automatically eliminate all controversies concerning incorrect dimensions. **After all, how can something be wrong if it isn't even there?**

7. Architect to place note on plans leaving selection of all wall and ceiling finishes up to the wishes of the builder. The same goes for color selections, hardware, electrical items, and plumbing fixtures. A general note on the plans stating that the mechanical contractor should furnish a comfortable level of heating and air conditioning should take care of that problem.

8. Architect and engineer to add a general note on the plans as follows: "Contractor shall build all work to comply with local building department. Any discrepancies during construction between the various building trades shall be worked out by the general contractor, **without disturbing the architect or structural engineer at any time."**

I also have other helpful hints for architects and engineers:

What to do if a client doesn't pay you. Simply print your plans on blueprint paper that fades rapidly in sunlight, or better yet, paper that self-destructs when exposed to daylight. Either the builder will have to pay your bill, or else do the construction at night.

What to do is the contractor calls you 12 times a day to ask you questions about your plans that he doesn't understand. Tell the builder that you will bill for the phone time spent in explaining the plans.

What to do when an incompetent contractor works from your prize-winning plans and the building turns out with a roof sag like suspension cables of the Brooklyn Bridge; the walls tilt like the Pisa Tower, and the colors are opposite from those you had intended. In desperation, you might sell tickets to tourists and denounce the work, but in all likelihood there is a 50-50 chance that people will actually like it better the way it is. Simply act like that was the way you had intended it. It will bring you future work and commercial success.

•••

In no other line of work is the professional hired to such a great extent based solely upon fee. A common discussion among structural engineers concerns the fee proposals given to various architects and developers. As a general rule, the structural engineer with the lowest fee will get the job. A certain project may actually cost the engineer money, but to get a foot in the door with a successful architect or wealthy builder; these projects are considered as investments. In another scenario, the engineer may be a little low on work at the time, and is willing to take the job just to avoid laying off employees. In no other profession are members so openly critical of others in their field. Unlike the medical profession, it is relatively easy to find an engineer who will be quick to criticize another engineer's plans and work.

A standard agreement, whether written or verbal, is common practice between the architect and the consulting engineer. It begins with the words: "you'll get paid when *I* get paid." Except perhaps book publishing, hardly any other business operates to such a large extent in such an unprofessional manner. I still remember the first time that I approached a bank loan officer for a short-term loan just after starting my business. I was getting a fair amount of new work, and it became necessary to hire some designers and draftsmen to complete the jobs on time.

The loan officer asked the standard questions regarding when I was to receive payment from the architect. His mouth fell open as I naively told him that my agreement stated that I would be paid whenever he was paid by his client. Could I put a time estimate on when I could pay back the loan, he wanted to know. Well, I explained, my client promised me that it shouldn't be too long before he receives some payments, and as soon as he did, he would pay me in proportion to the amount that he was paid. What if my client would not be paid at all, queried the bank officer. Well, I replied, I guess I never thought of that. Why don't you ask him, he said.

When I confronted my architect-client with the question, he replied, "You certainly wouldn't expect me to pay you out of my own pocket if *I* don't get paid, now would you? You just have to realize and accept that this is the way the business works. If you can't get along with it, I guess I'll have to find another structural engineer who can." 'Oh no, that's alright,' I gulped. "But how will I know when you get paid," I gingerly questioned. "Trust me," said he. "Trust me," I said to the loan officer. After putting up my house, my car, and everything else my wife and I owned, the bank "trusted me'-' and I finally got the loan. What kind of a business did I get myself into, I thought. Little did I know that this was just the very beginning of my problems.

Not long after this incident I opened my consulting structural engineering office, and started to become aware of the numerous problems that plague the industry. I immediately joined a small organization of fellow engineers in business like myself who were trying to improve the situation. I spoke at a state convention of structural engineers on fees, while holding a tin cup in my hand to illustrate our position in the giant building industry. I had an article published in a California building industries magazine that dealt with the consequences of low fees. Later I spoke to other groups of engineers on the same subject. Unfortunately, solutions to the problems were hard to find, and the interest gradually waned. It became impossible to justify remaining in a profession which did not seem interested in bettering itself and I finally decided to get out, as so many others have done before me. I became increasingly upset by the many building failures occurring almost weekly, knowing what probably caused the failures, and knowing how easily they could have been averted. It then struck me... how few people in the entire country have even the faintest idea of what is really behind these failures... and how few people ever hear anything about the architectural and structural engineering professions. Individuals I spoke with concerning non-ending liabilities, low fees, inadequate plans, and the causes as well as the great number of building failures were astonished at what I had to tell them.

Photograph on preceeding page:

April, 1981. Investigators from federal and local agencies search wreckage of a five story condominium in Cocoa Beach, Florida, trying to determine the cause of collapse. While crews were pouring concrete on the roof, the entire building collapsed, stacking floor slabs one on top of the other. 11 workers were killed.

–Courtesy of the Miami Herald

Chapter Six

Rain, Snow, Wind and Earthquakes

Many failures in buildings are caused in whole or part by natural elements such as rain, snow, wind, or earthquake. In general, building codes recommend the minimum design values. Of course, the proper code must be used, since there are local, city, county, state, regional, and national building codes, and the actual snow loads may vary drastically over a short geographic distance from one location to the next. This is especially true in mountainous regions where weather conditions can be severe.

Many people, however, forget that the recommended loads in these building codes are only *minimum* values, to be used with good judgment and revised upward as required by the design engineer. Official building codes can never account for unique situations unless they are properly augmented.

For example, if a roof is relatively flat as well as flexible, a buildup of rain water may cause the beams to deflect before the water can reach the roof drains. Hundreds of roofs have collapsed due to this oversight. Yet, no year passes without new buildings being constructed with roofs which are nearly flat! In a heavy rainstorm, as the roof portion begins to sag under the weight of the water, more liquid quickly runs into this low area, increasing the weight on the beam, which in turn causes more water to run into the area, increasing the weight even more, which in turn.... It requires little imagination to understand that, under these conditions a roof will soon collapse. Many of these kinds of failures are blamed on the weather. But this could easily be prevented by simply increasing the roof slopes, adding more drains, or by using a stiffer roof structure more resistant to deflections. In most cases all that is truly needed is the engineer's increased attention to reviewing the check list of "danger signals." There simply is no excuse for these types of building failures to persist.

Similar problems exist with regard to snow loads. Often, little attention is given to the possibility of drifting, which could pile snow several feet deep at vertical or sloping roof surfaces. Occasionally, a higher sloping portion of the roof is built adjacent to a lower roof level. As the snow melts from the higher roof, it slides off and piles up in great depths on the lower portion, leading to an overload failure. The engineer, in haste to meet a schedule, often glances over the building codes quickly, and simply designs the whole roof to support a uniformly distributed snow load of a constant depth. Although the letter of the law may have been followed, the spirit of good judgment certainly was not.

It is common practice, in many areas, for engineers to practically ignore wind forces entirely. A great many structures have collapsed due to a sudden large gust of wind. Again, the weather is blamed, while in fact, additional time spent on the project would have prevented the failure.

Few people realize that earthquake activity zones are located not only in California and the western states, but also in many eastern, southern, and central states, according to the nationally recognized Uniform Building Code. Many states, in addition to all the western states, have relatively high seismic risk zones within their boundaries. These include:

Alabama	New Hampshire
Arkansas	New York
Illinois	North Carolina
Indiana	Oklahoma
Kansas	Pennsylvania
Kentucky	South Carolina
Maine	Tennessee
Massachusetts	Texas
Mississippi	Vermont
Missouri	Virginia
Nebraska	West Virginia

Unfortunately, few, if any, buildings in the above listed states are ever designed to resist earthquake forces. All buildings in the western states require earthquake designs, and with very few exceptions, structural engineers in these states design buildings to resist these lateral forces.

It is interesting to note how a building is made earthquake resistant. An imaginary horizontal force is assumed to be pressed against the building, and the structural engineer then proceeds to design the elements of the building to resist this lateral force which tends to exert sideways pressure on the building. Of course, in practice, the engineering is more complex than this simple explanation and these designs require very involved intricate calculations.

It should be kept in mind that no building can be made completely earthquake proof (any more than it could be made completely fireproof). It can, however, be made earthquake resistant in varying degrees, depending upon the amount of money spent to strengthen the bracing systems. The degree of resistance is affected by the shape, size, materials used, and height. The importance of usage is always considered. For example, higher degrees of earthquake safety are built into hospitals and schools than are warehouses and industrial buildings which usually house fewer occupants. The amount of seismic safety is sometimes specified by the owner.

Only an actual earthquake can really test the adequacy of earthquake design and construction. The most important item in earthquake design is to be certain that all of the various building components are well tied together. After investigating the structural damage to many southern California buildings after the 1971 earthquake, it was learned that much of the damage was due to unconnected steel plates, tie straps, unbolted and unwelded connections, and other missing items; although all of these had been fully detailed on the original plans. This again points out the great need for better inspection and supervision by the design engineer.

The role of building departments in checking and approving the building plans has not changed greatly in the past several years. In all fairness, they cannot begin to check the myriad of details required for a major building. At the outset, that kind of inspection would require an increase in government workers, and even if these officials could do a complete examination, they would still not incur any responsibility or liability for the errors. Keep in mind that all building collapses of the past involved buildings that had been issued permits. They each were inspected and approved... yet they still failed. The core of the problem remains tied to the fact that the design architect and structural engineer *must* spend more time... a great deal of more time... and effort... on the plans and details. Again, they cannot be asked to do this without additional compensation.

Every year or two, many houses and smaller buildings are damaged, some beyond repair, due to high winds, tornadoes, and hurricanes. Occasionally, earthquakes also damage some of these structures. It is surprising how little extra it would cost to make a house or a small building resistant to these forces of nature. I estimate that these added costs would rarely exceed $200 per smaller house! All it would take is a little plywood, a few tie straps at crucial locations, and a few extra nails and bolts. These small measures can make a world of difference by increasing structural resistance manyfold. Here is a list of some suggestions:

1. Add extra anchor bolts from wall sill plates to the foundations to help keep the building from sliding off its supports.

2. Add vertical tie straps to tie down rafters to walls and walls to foundations. This will help keep high wind forces from lifting the roof structure off from the supporting walls and footings. (This is especially useful in tornado and hurricane areas.)

3. Provide horizontal metal tie straps where top wall plates are discontinuous.

4. Occasionally, short plywood panels should be placed at exterior walls which contain many windows or glass doors. This item, above all else, can add lateral stability to the entire building by providing needed bracing.

These suggestions should be considered a necessity in parts of the country where tornadoes, hurricanes, or earthquakes are common. Although the added cost is low, a small percentage of American homes are constructed in this manner Of course, if your home is sitting directly over the epicenter of an 8.0 earthquake, or directly in the path of a devaststing tornado or hurricane, little can be done. However, for the hundreds of thousands of houses and small buildings that are located adjacent to the major concentration of the force itself, the savings in lives, injuries, and property damage would be considerable.

Earlier I mentioned the problems associated with professional architects and engineers who feel their work receives "no respect." The problem extends also to the buildings themselves being unappreciated. This may sound strange, for what I mean is that a building's ability or inability to support added loads that may be attached to it should always be considered.

Building users and occupants should be educated in the support limitations of their structures. For example, many industries will carelessly hang hundreds or even thousands of pounds of heavy equipment loads from roofs without ever bothering to have a structural engineer look into its overall safety. A friend recently remarked that he knew of cases where many tons of audio equipment had been suspended from a roof by sound technicians without a second thought! It takes little imagination to realize that roofs are not normally capable of safely supporting this magnitude of weight, especially if they also happen to be supporting snow or rainwater loads. In these particular circumstances, a few inches of snow could easily cause a roof to collapse.

Photograph on preceeding page:

March, 1980. The collapse of a high school gymnasium roof in Antioch, California was attributed to the failure of precast, prestressed roof beams. No one was injured since falling fragments an hour before the main collapse gave warning. This event led to the precautionary closing of five other California schools of similar construction.

- Courtesy of The Oakland Tribune.
Photographer: Russ Reed.

Chapter Seven

Liability... "Let's Sue"

It is amazing to see how haphazardly so many structures are constructed in America. The contractor often takes many unnecessary risks which at the time might seem cost efficient but can turn into disaster. For example, a builder constructing a three story building with brick walls may choose to build the walls without any supports or bracings whatsoever. If you can visualize an unreinforced brick wall standing 35 to 40 feet tall, before any floors have been installed, you may appreciate the prayers of the contractor for calm and windless weather. If a moderate or high wind gust happens while these walls are being assembled, a lawsuit cannot be far behind.

Everyone even marginally associated with such a project disaster is brought into the lawsuit, especially if injuries or deaths occur from the toppling walls. The structural design engineer may even be blamed, although no agreement may have ever been made with the owner/developer to provide construction supervision.

As a general rule, structural engineers do not design or supervise bracing or shoring systems for elements of a building during the construction phase. The contractor is left to make decisions on how to construct the building, and whether such braces would be necessary to shore up the various portions as the building proceeds. The contractor's work is to produce the finished building as depicted by the prepared plans of the architect and engineer. With a schedule to meet, contractors are awarded the job to build the structure for a stated price, and often feel that they should determine the methods by which they produce the finished product. If contractors are instructed by design engineer on **how** to build the building, they can thereby lose the necessary control over cost and schedules.

Most design engineers shy away from accepting the work of designing or supervising temporary bracing and shoring. They feel that the small fee they receive for this work would not begin to compensate for the increased responsibility and liability. Since most failures take place during the actual construction period, the risks are indeed enormous. Besides, an engineer would also require a great amount of increased power by the owner in order to exert the necessary control over the contractor, leading to the question of ultimate responsibility during construction. Would the owner be willing to pay for this? If so, the savings in lives and injuries, as well as money due to a reduction in failures, accidents, and time delays would be immeasurable.

Whose responsibility is it when a construction worker falls from the roof during construction? The design architect or engineer? The contractor? The owner-developer? Or, perhaps the worker himself?

Landscape architects have been sued because trees they may specified for use in parkways caused concrete sidewalks to buckle, creating hazards for pedestrians. The claims state that the architect should have known that the tree roots would someday cause a problem. I have been sued, and was required to pay thousands of dollars to defend myself simply because a room air conditioner condensate leaked onto an adjoining sidewalk where a woman tripped and fell. I had received $300 as my design fee fifteen years earlier for the structural engineering of this small apartment building, never dreaming that I would have to pay several thousand dollars eleven years later due to a worn down air conditioner! As the structural engineer, I had nothing whatsoever to do with any mechanical devices, yet that did not stop the lawsuit.

In another case, I was sued when a carpenter installing roof joists slipped and fell from atop a one story building 300 miles from my office. Again, I had been hired to provide structural engineering plans, with no field supervision requirements in my agreement. Yet, it cost me thousands of dollars and many hours of depositions and meetings with an attorney to defend myself. In these types of lawsuits, it matters little who was actually responsible for the damages, since the party with the most insurance or financial ability is often the one who is required to pay the greatest share of the losses. This is commonly known as the "deep pocket" theory, since the one with the "deepest pocket" pays the most.

Justice in the building industry has taken a strange turn in recent years. People are not only held responsible for their own actions, but must find someone else to blame. If someone is injured or killed, someone must pay — regardless of whose actual fault it may have been. The parties with the most money or the largest amount of insurance will ultimately pay. This is especially true if there is some kind of an emotional appeal apparent to the court or jury. The sight of a young widow dressed in black, weeping at the sidelines along with her small daughter will always make a strong impact, and may assure that the party that can afford it will be required to pay, regardless of blame. The current line of public thinking, in which the "big company" is automatically believed "evil", while the typical citizen is assumed "good", has found its way into the court system. One of the problems is that the architect or structural engineer is often grouped as a team-player of "big business" through association, and the injured party may go to any lengths to try to collect the money, again regardless of actual blame. Unfortunately, the majority of the design professionals are not large firms and cannot afford large defense costs. Even if they are found to be faultless, it is rare that their legal defense costs can ever be recovered. Our laws operate on the premise that anyone can sue anyone else, and leaves it completely to the defendant to pay the defense costs, regardless of the justification for the lawsuit in the first place.

The "deep pocket" theory was evident in one of my projects which had problems with a leaky roof. The problem had nothing to do with the actual roof structure. In fact, even the architect in this case was not at fault in my opinion, since the contractor had revised the original architectural detail himself. Nonetheless, I was sued, along with the architect for many thousands of dollars. When it was discovered that the architect had no liability insurance, the case was dropped against him. Since I was the structural engineer and did have insurance, I was kept on as the defendent. After paying the insurance company attornies several thousand dollars to defend me, the case was finally dropped. The plaintiff had even admitted that the reason he maintained the suit was because I was the one with insurance!

The layperson may think that the architect and engineer need only carry professional liability insurance, similar to a doctor, or the typical automobile driver. Wrong! Although the design professional may carry this type of insurance, all available policies have very large deductibles, with a typical minimum deductible amount of many thousands of dollars. Worse yet, this deductible amount must be paid by the architect or engineer for legal defense costs, whether or not they are found to be responsible for the damages. This is like writing a blank check to the insurance company, good for any amount up to the stated maximum, to pay these defense costs each and every time a claim is made. This money must be paid, even if the insured has nothing to do with the charges. This is the reason why many architects and engineers do not carry any professional liability insurance. Without insurance, they are often released from the lawsuit early, while the plaintiff proceeds against those defendants that are insured. This is a fact of life in the building profession.

•••

In ancient times, before our current sophisticated technology came into being, builders would use the old "trial and error" method of construction. It consisted of three simple steps:

1. Build it.
2. If it falls down, make it stronger and build it again.
3. Keep doing the above until building stands.

Even during Babylonian days, the hazards of the construction industry were well recognized. In the Louvre in Paris stands an 8 foot column on which the Code of Hammurabi (circa 1950 B.C.) is engraved.

The Code is simple and direct:

If a contractor builds a house for a man, this man shall give the contractor two shekels of silver per ser (a unit of weight) as recompense.

•

If a contractor builds a house for a man and does not build it strong enough, and the house which he builds collapses and causes the death of the house owner, then the contractor shall be put to death.

•

If it causes the death of the son of the owner, then the son of the contractor shall be put to death.

•

If it causes the death of a slave of the owner, then he (the contractor) shall give the owner a slave of equal value.

•

If it destroys property he, (the contractor) shall replace what has been destroyed, and because he did not build the house strong enough and it collapsed, he shall rebuild the house at his own expense.

•

If a contractor builds a house for a man and does not build it so that it stands ordinary wear and a wall collapses, then he shall reinforce the wall at his own expense.

Apparently in Hammurabi's time little thought was given to stresses and strains, live loads, and building departments. Also, there was little chance of buck-passing. All responsibility was centralized to one man in the simple civilization of that day. Today, the collapse may involve the contractor, the architect, the structural engineer, the building inspector, and the city plan checker.

In my experience as a consulting structural engineer, I was also sued by a woman who had tripped on a small step 13 years after the building was constructed. The step was not even indicated on my drawings, since the plans called for a sloping ramp without steps. The builder had installed the step based upon his own decision and without my knowledge. A notorious Brinks robber appeared in the news headlines several years ago, having been captured shortly before the statute of limitations expired. Yet, most states have no statute of limitations for architects and engineers. Doesn't it seem strange that the engineer can be held liable so many years later, while an embezzler or bank robber can be free if not apprehended within a given number of years after the crime?

Several states which did have statutes of limitations for architects and engineers have had these laws reversed and found unconstitutional. Recently, the Wyoming Supreme Court reversed a state law which granted architects and others immunity from liability for building defects discovered more than 10 years after construction had been completed. The Court stated: "The statute in question is not a statute of limitations but is a grant of immunity from suit. This immunity is conferred only on a narrow spectrum of defendents. We hold there is no rational or reasonable justification for granting this immunity to this limited class of persons." In this particular case, the plaintiff had declared "There is no reason to protect architects...." Our laws are very peculiar indeed when statutes of limitations are acceptable for felons such as bank robbers, but unacceptable for design professionals.

The enormous amounts of energy and cash spent in defending against unjustified lawsuits considerably increases the cost of doing business in the architect-engineer profession. The result is that less time and money remains to actually do the job that is needed. The public eventually suffers by having poorly designed and constructed buildings in which to live, work and play.

Even the famous, well respected architects are not immune to these problems. For instance, Edward Durell Stone (1902-1978) was an architect who gained fame for many outstanding buildings, including the U.S. Embassy building in India, and the Museum of Modern Art in New York City. One of his outstanding achievements was the design of the $73,000,000 John F. Kennedy Center for the Performing Arts in Washington, D.C.

According to published reports, Stone's firm received $2,168,000 in fees (approximately 3% of the project cost) while incurring costs of $2,180,000. "Without a fee increase," he said, "I will have contributed six years of my life without compensation for my services or profit for my organization."

After Stone's death, his estate agreed to pay the government $248,451.94, thus settling a countersuit for alleged design deficiencies regarding roof leaks. The government agreed to pay his estate $223,000 in fees which had been withheld. The lawsuit, initiated by Stone to collect $295,000 in fees he felt were due, was answered by a government counter suit for $1,975,000, which was later reduced to $699,000.

Sometimes an error on a set of plans is not even an error. I know of one situation where the flair and style of a draftsman caused a great deal of extra expense in the construction of a 20,000 square foot drug store. This draftsman at times had written the number "5" to appear like the number "3," and a particular dimension on the plan for the length of many glued laminated wood roof girders was supposed to be "35 feet." The draftsman indicated it correctly, but the fabricator of the girders took it to be "33 feet." You guessed it... all of the girders were delivered to the job site short by two feet! The job was delayed, and a very expensive detail was created to add 2 foot lengths of wood to the girders.

Large government projects are not immune from structural failures either. Several Air Force barracks in the Midwest had extensive failures in the basement masonry walls. After rainstorms the earth surrounding the buildings became saturated and the increased fluid pressure caused the walls to buckle inward. Emergency bracing systems were rushed to the scene until the walls could be rebuilt — at great taxpayer expense. In these cases, it was later learned that the buckled walls did not even contain reinforcing bars which could have easily prevented this kind of building failure. Not long after my graduation from engineering school my first assignment at the U.S. Corps of Engineers was to review plans for government warehouse buildings which were several hundreds of feet long. I asked my supervisors how the buildings should be braced against wind forces. They replied that these "special designs" were really not required and that I should just check to see that there is some kind of a small brace "every so often." Even then I failed to understand the reasoning behind these lax structural requirements, little wonder that so many government buildings continue to fail.

Sometimes, it is possible to predict certain problems that are likely to occur during building construction. For example, on a three story major department store for which I did the structural engineering design, the lower story on one side of the building was below the outside ground surface. The masonry walls were designed to act as earth-retaining walls. To construct this building, the entire ground floor structure over the basement was required to be built first and I placed several large notes on the plans instructing the contractor not to place earth against the wall until this floor was in place. The contractor dismissed these notes as unimportant, and proceeded to place the earth fill. During a visit to the job site, I noticed a very unusual tilting along the 300 foot long wall. The problems this created with the remaining structure were substantial. None of the steel framings would fit beyond that point. The cutting, welding, and extensive reworking of all the remaining connections went on for weeks!

If one thinks that, under the present system, plans for buildings are adequate, just ask any contractor how much adjusting is done in the field during construction to get the building to fit together properly. Ask how many errors in the plans need correcting while the building is constructed. You see, the contractor is the "last chance" before the real damage is done. If the contractor does not find and correct the errors of the designers, the errors will forever plague the structure. But remember, the contractor is not a structural engineer, and if there are too few bolts or welds in a connection, only the engineer would really know it. Only the obvious "common sense" plan errors should be expected to be caught by the any contractor.

Following the San Fernando Valley earthquake of 1971, I spent a great deal of time inspecting many of the damaged buildings. In some cases, had the earthquake continued for only another half minute or so, much greater damage would have resulted. In a major post office which I had engineered, some second story precast concrete wall panels were to have been welded to the second floor. It was only after the earthquake had moved some of these panels several inches, almost completely away from their supports, was it found that the contractor had never made the welds at all. With another 30 seconds or so, and the second story walls would have become first story rubble. What is amazing is that sometimes it takes so little to make a building safe... several extra bolts, a few steel plates, a few welds — it would make all the difference in the world.

Chapter Eight

Computers, Common Sense, and Competency

A recent article appeared in newspapers throughout the country, charging that a serious crisis exists in the overall structural design of our long-span, column-free buildings. The article suggested that perhaps our desire for large open-space buildings has exceeded our ability to design and construct them safely. According to the author, too many architects and structural engineers, at our peril, are guilty of negligence that victimizes an innocent public with slumping on-the-job incompetence.

These are indeed serious charges. But I refuse to believe that most structural failures are due to the incompetency of design engineers. Competency is really not the issue at all. Graduate engineers are among the most intelligent, creative, and capable members of our society. Those that enter the field of structural engineering join engineering societies and are required to continue to take advanced courses and to keep up with current technology. As such, they have the dedication and a drive to succeed in their careers.

The problem, as I have previously mentioned, is with the collective building industry, the "system," rather than any particular group or individuals working within this industry. It is a system that fails to acknowledge the needs of the design profession; a system that does not recognize the importance of providing sufficient time or adequate payment for the various necessary services, and a system that does not understand the enormous responsibilities of the architect and structural engineer.

Consider that while in school, a student who might receive an average grade of 98% on all examinations during the university studies. The typical student also participates in other required activities and as an outstanding achiever may rightly be granted special honors upon graduation. This person's Bachelor of Science degree may be designated by the Latin phrases "Cum Laude" or perhaps "Summa Cum Laude." However, a 98% correctness factor in the field of structural engineering in the real world could readily mean disaster. It could mean that 2% of the floor or roof beams are undersized, or that 2% of the connections are inadequate. Not bad — unless you happen to be standing under that particular beam when it collapses!

A grade of 100% is required in the real world — nothing less will do. Nothing less should be tolerated. However, we all must be realize that no person on this earth is actually capable of doing 100% work all of the time. Does this mean that the person is incompetent? Of course not, but what it does mean is that special precautions must then be taken in order to achieve total accuracy on a completed engineering project design. Here is where a great problem arises, and no one seems to take the time to grasp it.

When we consider the extreme complications; the hundreds or even thousands of decisions required to accurately design an average building, we find that the only real way to achieve the goal of 100% accuracy is to do the design, check it, and then have a second qualified person or team re-check the work. This verification should be done in addition to any independent coordination or checking that is needed to mesh the plans with the other building trades such as architectural, mechanical, and electrical plans and details. Of course, all of these activities demand additional time — and money. You see, one engineer performing design work cannot do 100% accurate work the first time through. However, checking and reviewing alone and with others can eventually yield the perfection required. Believe me, there is no other way.

In recent years, there has been an enormous increase in the complexity of structural designs. Many of the new design methods are highly theoretical, having been recommended by engineering committees and detailed studies from major recognized universities. In some instances, the new requirements are so complex that the typical practicing engineer cannot even begin to comprehend the complications of these new techniques. This may lead to errors during interpretation as well, resulting in incorrect structural designs. In many cases, computer analyses are required.

Because of these changes, the need for good common sense is needed more today than ever before. It is so easy and habit forming to become dependent upon computers for answers to complex problems. If the human input to the computer is not continuously spot-checked, dangerous errors could likewise result. Computers cannot replace common sense. If, for instance, a computer printout is completely wrong, it will require good, experienced judgment by the design engineer to locate the error. In one case the plans for a multi-story steel frame building were completed and submitted along with the computer printouts to be checked by the building department. It was discovered that the computer analysis was based upon each steel frame carrying only one-half the actual required loading! Had the programmer's error not been caught prior to construction, sagging and twisted floors, if not a complete structural collapse, would have resulted.

A perspective on computers: during a lecture, a nationally known and respected author made the statement that "computers are today where writing was when it began 6,000 years ago!'' A profound thought to be sure, but as with all computer usage, the programmer (the engineer in this case) must be experienced and knowledgable in entering the proper information, and must always be on the alert for mistakes.

Once, while checking a school project, I was responsible for reviewing the concrete foundation pile system intended to support a large gymnasium building. I discovered that the piles were only deep enough to support one-half of the actual building weight. It seems that the pile capacity charts furnished by the foundation engineer in his report were based upon ''kips,'' with one kip being equal to 1,100 pounds. The design engineer had incorrectly assumed that the chart was based upon ''tons'' of capacity, or 2,200 pounds. The net result was a major overload. Again, a failure was averted, one that would have resulted in great damage and settlement of a building. The designer thanked me for finding this error, and I felt that I had earned my pay for that day.

The architectural and engineering schools are doing an excellent job of turning out well-educated graduates. With all of our new computer technology, the design professionals today have all the necessary knowledge and tools to do an excellent job at designing buildings — including the complex calculations for building modern span space frames.

Chapter Nine

Perspective

Let's look to the past for a while, just to get a better perspective on the changes that have taken place since some of the famous classical buildings were erected. Consider one of the world's most famous architectural wonders — the Leaning Tower of Pisa, which also happens to be the grand-daddy of all foundation mishaps. I might also refer to it as one of the all time "greats" in building failures — a failure that made good!

The Leaning Tower is a bell tower, and as such, is a relatively small structure compared to some of our modern high rise buildings. If it were designed and built today, it might have to meet a time schedule of about three months for completion of the plans, and another six to nine months for actual construction. In other words, from start to finish, less than a year would pass. In history, the Leaning Tower actually took 200 years to build!

The story goes like this: The tower was begun in 1155 to compete with the bell tower of St. Mark's in Venice. Before it had reached 40 feet, it began to tilt noticeably. Founded on a deep volcanic ash, evidently of uneven consistency, it began to settle on an angle. The architect, in an attempt to correct the condition, built the next levels of collonades more nearly perpendicular. However, the tilting persisted. Discouraged, he left the job. Sixty years later, another architect took on the challenge, but after finishing the fourth arcade, he also quit. A third architect, William of Innsbruck, attempted to correct the condition by varying the lengths of the fifth and sixth arcade columns, however the settlement continued and he also gave up in despair. Nearly one hundred years later, the fourth and final architect, Tommaso Pisano, completed the sixth balcony, and built the belfry in a more vertical position. The building was finally completed in 1350, nearly 200 years after it was begun.

The marble walls of the tower are 13 feet thick at the base, narrowing to about 6 feet toward the top. Today, this tower would probably be constructed of reinforced concrete varying from only 12 inches at the base to 8 inches at the top, a reduction in materials of about 1200 percent! From the subsequent great weight reduction, it is possible that the settlement would not have occurred and the building's fame might not have been so widespread. Now, 600 years later, the Leaning Tower of Pisa continues to attract tourists at the rate of 350,000 per year. It is strange how things have changed. Today, a delay of 30 days in a construction contract is enough to give everyone involved a case of ulcers.

Not too many years ago, when a structural failure occurred, it was not hushed up. The event was publicized and the causes for the failure were widely distributed both to the public and the professions. This information was used and thoroughly studied by design engineers to assure that similar types of mistakes would not be repeated. In contrast, today's active climate of expensive lawsuits, considerations of liability and possibilities of litigation preclude the publicity of these kinds of technical data. Today everyone loses — the engineers, architects, owners, contractors, and the general public. Because of this insatiable paranoia of legal hassles, a desire for secrecy has developed which leads to repetition of the same types of errors in new buildings throughout the country. The suppression of engineering information on building failures is especially felt after a major catastrophe such as an earthquake, snowstorm, or hurricane. It is difficult to get anyone involved with a damaged project to freely discuss the causes of failure. Recently an international organization which collects and shares data on structural failures has agreed to formalize a failure-reporting system and is currently pursuing university repositories for these data. The system is named the Engineering Performance Information System (EPIC), and is primarily intended as a teaching aid. However, it will also be a resource to design structural engineers, for practical information and background data on litigation. The U.S./European group met to work out a computer program which will standardize the reporting format. Present-day information is so spotty that it is often impossible to distinguish between cases of physical failure, such as a sheared bolt, and procedural failures caused by the structural designer.

In my opinion, our own government has been contributing to the problems of poor and unsafe building construction. In several states, engineering organizations have attempted to improve their profession and the quality of construction by trying to eliminate situations where design engineers are chosen and hired solely on the basis of their fees. In numerous cases, the state and federal governments have filed lawsuits charging these organizations with price-fixing and restraints of free trade in violation of the antitrust laws. In other words, the government is encouraging competitive fee bidding in an industry where low fees and fee shopping is already far too prevalent. This seems to be the current law of the land, and I would like to encourage these organizations and societies to either work to have the unfair laws changed, or take a different approach towards improving the conditions within the construction field.

Concerning the problem of liability — the proper fee to charge for a project which takes relatively little time, but incurs great amounts of liability has always been a dilemma for engineers. Let's say that a structural engineer is commissioned to design a small portion of a large repetitive building project; for example the design of a series of typical roof joist, floor joist, and a beam for one unit of an apartment building. The engineer might spend three days on these design plans. There may be 125 identical apartment units within the project, representing a total construction cost of several million dollars. Should the engineer base the fee on three days of work, or on the consideration that the design is being used in a multi-million dollar project? If there are any problems during or after construction, the engineer could be exposed to an enormous liability on such a large project. I have always felt that the engineer's fee should be separated into three parts:

1. A fee based upon actual design time spent;
2. A fee for supervision of construction, and
3. A fee to cover the responsibility and liability incurred.

Concerning the payments made to an architect and structural engineer — one of the most frustrating situations encountered by the design professional is the time delay in obtaining payments from a client. Every architect and engineer that I know has complained of this. After closing my consulting office, I had to wait well over two years before the last payments for my work were finally made.

Concerning design work that is done for a developer where a decision was made to not build the project — if, for example, the project financing to build could not be obtained, or one of the financial partners backs out, the attitude of the developer to the architect or engineer is often, "Why should I pay you for a set of plans that I can't use?"

I once attended a meeting with the developer and the foundation engineer of a multi-story apartment building. I was to do the structural engineering for the project, including the designs for the footings in keeping with whatever recommendations the foundation engineer included in the soil report. This engineer, who was also employed by a testing laboratory, felt that the building should be supported by deep reinforced concrete driven piles. The soil was sandy, and in his opinion there was a slight possibility of liquifaction during an earthquake. (Liquifaction is a phenomenon which occurs when underlying soils of certain materials are subject to sudden settlement during a major seismic event, creating the possibility of sinking and damage to the building.) The developer listened, then made some quick estimates of the additional costs involved — somewhere in the neighborhood of $40,000. The discussion that followed between the developer and the foundation engineer clearly left me with the impression that if the engineer insisted on recommending the deep pile foundation system instead of the more conventional shallow footing system, he would probably not be receiving payment for the work that he had done. He had already spent much time on the project, drilling for soil samples at great depths, analyzing and testing these samples in a laboratory, drawing up charts and graphs, and preparing the detailed report with his recommendations. I returned to my office, and heard nothing more about the project. Two weeks later, a letter arrived from the foundation engineer, this time a revised report, stating that he *now* felt that the less expensive shallow footing system was satisfactory! (I wonder what changed his mind?)

One of the more serious problems facing the owners of the typical small to medium structural engineering firm is the feeling of a loss of control. Keep in mind that these smaller offices represent the majority of structural design firms throughout the nation and produce the largest dollar volume in buildings. The lack of control comes about because the office cannot perform an adequate job within the time schedule or fee scale demanded by the client and remain in business. The owner of an engineering firm is personally responsible and liable for each and every line and word that appears on the drawings, which are required by law to signed and sealed. Even if an owner could hire only the "98 percenter Magna Cum Laude" graduates mentioned earlier, (a little impractical), the principal would still have to be concerned over the 2% errors that may slip through. Such mistakes in plans could be enough to cause the loss of a license as well as one's entire personal property if a disasterous failure should occur, no matter how small the error on the plan. Large structural engineering firms, of which there are very few, have a sufficient volume of business, at slightly higher fees; more financial help available from banks, and are consequently able to afford the higher insurance premiums to avoid a financial disaster should anything go wrong. This fact however does not necessarily make the work of a large firm to be any better or safer. The collapse of many large structures designed by large firms provides testimony to that fact.

Many structural engineers with whom I have spoken over the years express a wonder at how some buildings ever manage to be built at all considering the poor plans produced by so many offices. Worse yet, it appears that some contractors are in competition with each other to see who can leave out the most connections! In earthquake country, many extra straps, angles, bolts, and nails are needed to tie the building elements together. These are always detailed on the structural plans, and should be installed by the contractor during the structural portions of construction. It is always a shock when I visit a project after the contractor tells me that his structural work is complete, and at the job site, usually behind some boxes or a rubbish pile, lies a large pile of unused straps, angles, and bolts!

It is not possible to discuss safety in building design and construction without at least mentioning the excellent work of the California Office of the State Architect Safety Section. This office began as the Division of Architecture, following the 1933 Long Beach earthquake which caused widespread destruction.

Many buildings collapsed, including schools. Fortunately, classes were not in session at the time, or many children might have suffered injury and even deaths. The office was established after a new law, passed by the state legislature, stated that all public school buildings must be designed and constructed to meet strict earthquake safety requirements. Since that time, over 50,000 school buildings have been built in California, all meeting these rigid safety measures. Many of these have since experienced several major earthquakes, with little if any damage. The plans are designed by private architectural and engineering firms to conform to the rules and regulations. Then, all plans are submitted to the state offices, where they undergo a strict plan check by licensed structural engineers. Following plan corrections by the design professionals, the buildings are constructed by contractors in strict compliance with the plans. During the entire construction period, there is an on-site building inspector who continuously checks all operations to insure that all of the work conforms to the plans in every detail. Also, the architect and engineer must make periodic inspection visits to the site to review the work, and they must submit verified progress reports at scheduled intervals. Any work that does not comply with the plans is immediately removed and redone. The result of of this procedure is that no other construction that I am aware of even comes close to this degree of safety, regardless of whether one is considering earthquakes, wind or just vertical floor and roof loading. The excellent record over the past 50 years bears out this fact. Following the San Fernando Valley earthquake in 1971, in which several major hospitals underwent great damage, even stricter regulations were drawn up to apply to all new hospital construction in California, now watched over by the same agency.

It is unfortunate that conventional buildings cannot be built to the same standards as California schools and hospitals. There are two basic reasons for the extremely high quality and strength of construction:

(1) a thorough review and checking of the architectural and structural engineering plans, and

(2) careful supervision and inspection during the actual construction.

Here is proof that a multiple check approach can work. The cost of the additional work and time spent by the state in this case is billed to the school districts, who pay this extra fee directly to the state agency for the review and checking involved. The money, in effect, comes from local and state taxes. In the case of hospital construction, these added costs are paid directly by the hospital owners. This is all satisfactory and seemingly justified because the public welfare is directly affected. However, I do not feel that government checking agencies should be expanded and involved in conventional building construction. The increase in the number of civil service employees throughout the country to establish such an agency would be astronomical and difficult to justify. After all, the additional checking and supervision of non-public construction should the responsibility of the architect and structural design engineer, and rightly so. But this service cannot be performed without provisions for a higher fee.

How can we, then, assure safety and solid construction within the existing system? It will take much dedicated effort by the committees of architects and structural engineers. Up to now the problem has been of low priority. So much attention has been given to the study of new technologies, materials, computers, and similar items, that this most important issue of all has been practically ignored. What happens is that when the business-part suffers, everything else suffers accordingly. Some day, the professions will wake up to this fact.

Some suggestions for improvement:

The owner-developer could insist, contractually, that a specified portion of the architect's fee must be paid to the structural engineer consultant, or directly to the engineer. By doing this, the owner-developer could control the fee that applies directly to the safety of the structure. Presently, developers may be paying an above average fee to the architect with only a small portion alloted to the engineer, thus leading to an inadequate, unsafe structure.

The construction lender, as well as the permanent financing lender could insist on approving the fees to the architect and structural engineer prior to making a commitment on the loan. To support this, the lender would also engage the services of an independent structural engineer, hired for the specific purpose of reviewing the design engineer's calculations and drawings. These licensed engineers would share in the responsibilities and liabilities along with the designers. They would be specially certified in particular fields of construction, such as high-rise structures or long span space frames over auditoriums.

Perhaps, in this way, the owner-developer could share in the liabilities of a building for an infinite number of years after construction, as do the architects and engineers. After all, they are the ones who benefit from the inadequate fees paid to the design professionals. They are the ones who actually select the design team, and so it is only fair that they share the liability. Presently, a developer can build a large building cheaply and poorly, sell the building soon after construction, and walk away with the profits. Should any problem develop in the future (indirectly the result of the bargain fees paid), the architect and engineer are held fully responsible — forever.

Apparently there has not been sufficient publicity given to the great many building failures and near disasters to stir public concern. We have been very lucky so far since so many of the major failures have occurred when these structures were unoccupied. For example, nearly all of the major auditorium roof collapses have occurred when they were not in use. Had the roof caved in during a major sports event or convention meeting, many thousands of injuries and lost lives would have occurred.

In any discussion on building failures, it seems appropriate to mention some problems which are really not serious enough to cause a collapse. They may be annoying, however, and deserve more attention from the designers. In many areas, these problems can be easily minimized if the effort is made. Here is a partial list:

CRACKS IN CONCRETE AND MASONRY.

By far the most common cause of cracking in walks, slabs, beams, and walls, is shrinkage of the material itself as it ages. It takes an experienced engineer to actually determine if more serious problems are present. Hundreds of technical papers have been written on this subject. Normally, although the cracks may be unsightly, they do not create any safety hazard. Shrinkage cracks may be reduced by using special expansion and control joints to allow the cracking to take place at predetermined locations.

PROBLEMS WITH WOODEN MEMBERS.

The basic problem with wood is that it is non-uniform in texture, and therefore unpredictable. No two trees are alike. Differences in moisture content, knots, grain slope, splits, and checks can cause a nightmare for anyone designing or working with the material. As the lumber ages, it slowly loses (or gains) moisture to match its final environment. As it does so, it may twist, sag, shrink, or split. Depending upon its location and use in the structure, visible problems could develop in various degrees. Care must be exercised by the architect and engineer in its use. Due to its low cost compared with other materials, wood is sometimes used in locations where another material, such as steel, would be more suitable. In bolted connections, for instance, it is easy to forget the effect that shrinkage may have in causing dangerous splits in the supported beams. Problems can be avoided by careful detailing of connections, and by careful analysis in determining long-term deflections.

"BOUNCY" FLOORS.

The problem with "bouncy" floors is common with all types of materials, whether concrete, steel, or wood. Although it may be annoying to have the floor vibrate or bounce whenever someone walks on it, this usually has little effect on the structural safety of the building. In all cases, the problem could have been avoided by using larger and stiffer structural members. Unfortunately, increased costs are a factor, and there is strong hesitance of many owner/developers and engineers to provide more than the minimum member sizes.

SAGGING CANOPIES OR FASCIAS

Sagging canopies can be one of the most annoying problems of all. No owner wants to accept an obvious sag or deflection on the front of a new building, even though the engineer tells the truth in explaining that there is no structural danger. There is really no excuse for this obvious mistake. The engineer must realize that the appearance of the building exterior is of primary importance and determining the structural member sizes should be influenced by appearance as well as strength in this situation. In doing so, the engineer must use all expertise and knowledge of materials, shrinkage, deflections, and connections to assure a perfect facade.

Owners and developers are well experienced in demanding precisely what they want from architects and engineers. They are people in business. They know that it is easy to intimidate design professionals who, as a rule, are not as business orientated. The owner or developer is basically concerned with construction costs, financing, projected incomes, and future profits. These are an owner's priorities. The design professional, on the other hand is usually more interested in the aesthetics of the building, such as its shape, materials, and colors, as well as technical aspects such as beam sizes, live loads, and bracing. When it comes to business, it's no contest.

Often, the structural engineer is coerced into specifying items in the plans that he may not completely agree with. For example, a developer may insist on the use of 2' x 8' floor joists, when 2' x 10' joists would increase the cost of the floor system by only about 5 cents per square foot, and could assure a rigid floor that will not bounce or sag under normal conditions. Compared to the overall building cost of possibly $60 per square foot, this five cents is negligible. Sometimes the architect will apply pressure on the engineer to use a hidden connection for a beam. Again, when the connection fails, the engineer will almost certainly be held liable. Many problems with structures have developed simply because the engineer gave in to such questionable judgments. Although the owner and the architect have the greatest overall control of the project, they are not structural specialists.

Many times, when a contractor's bid is presented over budget everyone looks for places to cut costs. As usual, no one wants to reduce appearance items, so hidden ones, such as the structural frame are chosen for modification. Often, the structural engineer is accused of overdesigning the framing members and wasting the client's money. The owner may even assign the plans to another engineer to confirm and check over the plans to determine if any costs could be further reduced. The second engineer, may often be eager to make a good impression on the owner in hopes of getting future work and may well present a list of some assumed structural errors, as well as items that could be legally reduced in size. These money saving ploys usually result in very stormy confrontations with strong professional disagreements.

The outside engineer, called in to review the plans can emerge in a favorable light to the owner, since a third party consultant can nearly always be able to cut some of the costs. (There has never been a set of plans that cannot be cheapened in quality — sometimes to a dangerous degree.) The motivation is obvious, an outside engineer can expect future work from the owner and/or the architect. Of course, he may be despised by the original engineer — but that's life!

The engineer who prepared the original structural plans may afterwards appear incompetent to the owner, who may then feel that the engineer has wasted his money. ("I told you I wanted a cheap job, didn't I?") The engineer may also be an embarrassment to the architect, and falls into a common no-win situation of deciding to (1) justify the original engineering design or (2) revise it, reduce the cost and admit to the criticisms of the outside consultant. Failing to do either of these, the engineer will probably not be paid the full fee from the owner. Revising the work by reducing member sizes may result in a building with structural weaknesses beyond those limits that were originally intended — leading to liability problems in the future.

The architect is also somewhat uneasy throughout the whole procedure, after all, he was probably the person who selected or recommended the design engineer in the first place. The owner's dissatisfaction reflects on the architect, who is also concerned about receiving a full fee when the building budget is finally approved. Most likely the architect will side with the owner against the original engineer, though not too vociferously.

The owner cannot lose, no matter what happens. Through an engineering re-evaluation some costs are bound to be reduced. The owner may require that if any revisions become necessary, to reduce the original cost estimates, these should be provided gratis by the architect and engineer. Also, the owner-developer might also outright refuse to pay for the charges; if the cost of the building so far exceeds the original budget that the owner cannot afford to build, he could even sue the architect for any payments already made. Or else, the architect might agree to completely redesign the building at no additional cost. All things considered, this is a very tense situation.

Sometimes, the original cost estimates leave a great deal to be desired. Once, when a one-story building which I engineered came in over budget, I was directly accused of overdesigning the roof. I clearly showed that even if the cost of the entire roof structure was deleted, leaving the building open to the sky, the building would have still exceeded the budget estimate!

Because of the present system of low fees and the lack of time given to do a thorough job, it is a commonly accepted fact among many architects and engineers that no set of plans should be expected to be perfect; that a certain amount of extra cost should be anticipated by the owner or contractor during the construction period; that after construction there may be a few places where a fascia or canopy may sag a little, or a few areas where the floor may be a little bit bouncy. After all, (this reasoning goes) considering the fees paid, and the rushed schedules, how could an owner expect anything more? And certainly no architect or structural engineer should be held liable for any damages after all of these considerations.

Unfortunately, for the design professionals, neither the owner/developer nor the contractor will usually accept this line of reasoning. The owner continues to feel that a fee is paid for a professional service, and will attempt to get that service even if it has to come out of the architect's and engineer's hide! After all, they agreed to accept the fee, didn't they? All this is nonsense! There really is no excuse for haphazard, sloppy, unchecked work. But, if the building designers are not going to give a complete service, for whatever reason, by all means they should at least be honest about it up front, and state what they are furnishing — unchecked work with the possibility of errors, discrepancies, and omissions. "This, in effect, is what you get for this amount of money. If you want a complete, thoroughly checked job, including supervision of construction, it will cost you (so many) dollars in addition. What do you want?" This should all then be put into writing. It may not help to keep the roof from falling, but at least everyone will know where they stand from square one!

What about official state building departments? Do they do much good, or are they just a way to collect more hidden taxes to keep the evergrowing population of bureaucrats in office? I am licensed in several states as a professional engineer, and have performed structural design services for buildings in various places during a fifteen year period. Most of my on California buildings received a relatively thorough plan check by engineers working for the local building departments. Inevitably, there were corrections to be made on my plans prior to receiving an official building permit. However, what has always surprised me was that I rarely, if ever, received a plan correction from any state other than California. Those plans were certainly not any more perfect than my California plans, yet apparently, no one bothered to check the drawings in these other states. This, is quite disconcerting. Here we have all this construction going on throughout the country, and building departments don't seem to even bother checking the structural drawings. In other words, a building permit in some areas does not really have very much to do with the structural safety of the building. In no way can these plans be perfect, and I again emphasize that throughout my career I have yet to see a set of plans free of errors, discrepancies, and omissions. The entire industry evidently operates on the hope that the errors will be small and unimportant, or that they will be noticed and corrected during construction. But what if they aren't? What happens then? What kind of professionalism is this?

One constant problem which plagues architects is whether or not a licensed architect is actually needed for the design of some buildings. Different states have different interpretations, and their laws vary accordingly. Many developers do not bother to hire licensed architects for their plan preparations at all. Instead, they use only draftsmen. Depending upon the size of the building, this practice is perfectly legal in many places. As a rule, large buildings are not affected, the larger architectural offices do not seem to be concerned over the problem. However, it is of major importance to most smaller offices. Unfortunately, they do not seem to have the clout to make changes in the laws, or even to see that the present laws are properly enforced.

The professions of architecture and structural engineering are extremely vulnerable to wild swings of the economy, high interest rates, and recessions. When the the overall economy is weak, few manufacturers think of expanding their plants. During these times not many developers can afford the high interest rates that must be paid to build housing. Even if they could afford it, the buyers could not. The very first industries to feel the impact of an economic downturn are those involved in planning future construction. To a large extent architects and engineers can actually be used as a guide to predicting the future economic conditions. When they are busy preparing plans, it is a foregone conclusion that construction will be busy for months to come. When construction is at a high level, all the affiliated industries can be assured of good business activity. Buyers of new homes and apartments will need furniture, carpeting, and landscaping. Consequently, thousands of companies that employ millions of workers can expect good times as unemployment decreases. The same is true for commercial construction. New shopping centers will be needed adjacent to the new housing areas. Office building construction usually marks the start of additional employment in many businesses, such as office equipment and office furniture.

Conversely, when architects and engineers begin to lay off their employees for lack of work, one can predict a coming downturn of the economy.

The business of designing buildings is unique. No other enterprise has such lengthy periods of liability after completion of its work. In this age of consumerism, few seem to accept responsibility for their own actions. Whenever an accident occurs, the injured party immediately looks for someone else to blame. Due to the numerous complexities and the very nature of the building industry, it is often extremely difficult to determine where the blame lies in accidents or building failures. The following list demonstrates the various types of circumstances which have lead to the many recent lawsuits against architects and engineers.

1. Someone trips and falls over a step in a ten-year-old building, claiming it was improperly marked.
2. An air conditioner unit leaks, someone slips and falls.
3. A roof drain gets clogged. The ponding rain water overloads the roof, causing it to collapse.
4. A carpenter mistakenly drills oversized bolt holes for a beam support connection. The bolt later slips, causing the canopy to sag.
5. Faulty fabrication of a glued laminated wood roof girder causes a roof failure after a heavy snow storm.
6. A welder forgets to weld a joint which is hidden from view on a steel frame which braces a large wall.
7. A carpenter accidently places a wooden beam upside down with the knot at the bottom instead of at the top. The beam cracks and fails years later.

The list of possible problems is endless. The wait to collect the fees are likewise endless. I recall the days when I yearned to own a hot dog stand. I could see myself selling a sandwich and immediately hold out my hand to collect the pay. The customer would gulp down the snack, walk away, and I would then be free! What a joyous feeling that would be today!

Photograph on preceeding page:

July, 1981. The collapse of the walkways at the Hyatt Hotel in Kansas City killed 111 persons and injured scores of others. It appears that connections at the supporting steel rods failed.

–Courtesy of The Kansas City Times.
Photographer: John Spink, © 1981.

Chapter Ten

An Open Letter to Architects and Structural Engineers

Shame, shame, shame!

You ought to be ashamed of yourselves! With all your education, knowledge, background, creativity, and effort, you do a disservice to yourselves as well as to the public. Whether you are principals of your firms, or employees, how can you possibly be content with the way things presently exist in your profession?

To employees — are you really happy making less than the construction workers who build what you design for them? Has it all been worth it? — the years of studying, the training, the struggle to get where you are? Don't you feel a little bit uneasy about the moonlighting which many of you do to make ends meet? Are you really helping your profession by doing this? Does your employer know? Wouldn't it be nice to be able to live as a professional, with recognition beyond name only? Have you thought of organizing a professional union? It is certain that you will to have to soon do something on your own to better your position, no one will hand it to you. Possibly by demanding better salaries, you could cut down on your moonlighting work and help the entire profession. Or perhaps you might think twice about starting your own business in architecture or structural engineering until you are prepared to charge large enough fees to do a professional and thorough job. Presently, the situation is such that many of you open your offices, content to earn slightly more than you could as an employee, without the foggiest notion of the responsibilities and liabilities you undertake, and with little knowledge of the pitfalls of the business-end of the profession. You immediately fall prey to developers who tempt you with carrots to get you to do projects at extremely low fees. If you can't change things for the better, perhaps it is time that you should get out of this profession!

To employers... Where is your sense of priority? How can you, in your societies and organizations, devote such a large and disproportionate amount of time to technical and aesthetic studies? What about the business-end of your profession, which is today so sadly neglected? When are you going to discover that the elevation of your profession to the distinguished, respected levels that it deserves will not come about until you are actually paid as a professional. Come now, you are in the unique position as the envied builders of America's future cities. What you create is destined to be around long after you have gone. You hold all the necessary creative power in your hands and wondrous imagination. Now don't blow it!

Can you honestly encourage your sons and daughters to enter this field, and feel right about it? Not many of us could. You might tell them that if they honestly enjoy architecture and engineering, they would make excellent hobbies, but if they desire to enjoy a successful well-paying career and be able to sleep well at nights, they would do well to choose some other profession. Not that some design professionals don't do very well — I venture to say that 5 percent of them make a very good living. But the other 95 percent live in mediocrity at best. Oh yes, some that stay in the field manage to do well if they are fortunate in having some good outside investments. But as far as making it in the profession itself — well, as I said earlier, it's a good hobby. But then, you are the "professionals" — why am I explaining it to you?

Of course, I am not able to spell out a step by step procedure that will end all your problems. I can however encourage ways that you can get the lenders involved on **your** side of the building industry. I can also tell about ways to put more teeth in the power of various state building departments — by having them require outside certified plan reviewers. I can likewise discuss the possibilities of your professional organizations running consumer ads in newspapers and magazines, and on radio and television; to educate the public as to what you do or don't do. Perhaps once the public learns, they may apply pressure to all those concerned in the building industry. But what you really must do above all else is dedicate some effort to the problem, and stop pretending that it does not exist.

To architects — you are the primary professionals that most often deal directly with the owner or developer, therefore you assume the basic responsibility for demanding and negotiating proper fees. You are also responsible for determining that the portion of your fee is to be paid to your consultants, including the structural engineer, and guaranteeing that this portion is sufficient for the work done. You see, when there is a structural failure, for any reason, you sink with the rest of the ship, whether or not you are personally to blame. When a beam connection fails, you can be certain that you will be sued. It is basically up to you to determine an adequate time schedule for the preparation of all the plans. Don't schedule the impossible or impractical. It is likewise up to you to shop around for structural consultants who charge low fees. However, you will be hurt by it, along with your client, if not now, perhaps some years later when you least suspect it.

To the structural engineers — I need not tell you about the importance of checking and re-checking your work — and then check it again! You already know of Murphy's Law (and its variations) concerning things going wrong at the worst possible time, or about the most important dimension on the plans having the greatest chance of being omitted, or of the design errors in any calculation being the one which could cause the most harm, and so on. Be strong. Stand up for your rights. Be a true professional in every respect.

To all professionals in the building industry — you must know as well as I do — nothing is going to change, unless YOU do something to change it! I wish you luck! You're going to need it!

Postscript

I am fully aware of the fact that many professionals other than architects and engineers feel underpaid and misunderstood. They, too, wish more respect as well as better treatment from their clients or employers. Life, of course, is not always fair. Income levels are not always commensurate with educational, experience, or intellectual requirements. One needs only to look at the teaching profession to see the truth to this condition. It has often been said, ''Unrewarded genius is almost a proverb. Nothing is more common than unsuccessful men with talent.''

It matters little to me whether or not design professionals receive their ''share of the pie.'' What does matter, however, is whether or not the safety of the unsuspecting public is jeopardized by the unconscionable methods used by the design and construction industry to run its business. It is the desire to publicize these facts that has compelled me to write this book. It is my conviction that only through this type of exposure can meaningful progress be forthcoming.

I ask that any comments, suggestions, or ideas concerning the views expressed in this book be forwarded to me in care of the publisher. Perhaps through this forum, steps may eventually be taken to improve the situation for all of us.